FORSCHUNGSBERICHTE AUS DEM LEHRSTUHL FÜR REGELUNGSSYSTEME

TECHNISCHE UNIVERSITÄT KAISERSLAUTERN

Band 14

T0140808

Forschungsberichte aus dem Lehrstuhl für Regelungssysteme

Technische Universität Kaiserslautern

Band 14

Herausgeber:

Prof. Dr. Steven Liu

Yun Wan

A Contribution to Modeling and Control of Modular Multilevel Cascaded Converter (MMCC)

Logos Verlag Berlin

Forschungsberichte aus dem Lehrstuhl für Regelungssysteme
Technische Universität Kaiserslautern

Herausgegeben von
Univ.-Prof. Dr.-Ing. Steven Liu
Lehrstuhl für Regelungssysteme
Technische Universität Kaiserslautern
Erwin-Schrödinger-Str. 12/332
D-67663 Kaiserslautern
E-Mail: sliu@eit.uni-kl.de

Bibliographic information published by the Deutsche Nationalbibliothek

The Deutsche Nationalbibliothek lists this publication in the Deutsche
Nationalbibliografie; detailed bibliographic data are available
on the Internet at http://dnb.d-nb.de .

ISBN 978-3-8325-4462-1
ISSN 2190-7897

Logos Verlag Berlin GmbH
Comeniushof, Gubener Str. 47,
10243 Berlin
Tel.:   +49 (0)30 / 42 85 10 90
Fax: +49 (0)30 / 42 85 10 92
http://www.logos-verlag.de

A Contribution to Modeling and Control of Modular Multilevel Cascaded Converter (MMCC)

Beitrag zur Modellierung und Regelung von Modularen Kaskadierten Multilevel-Umrichtern (MMCC)

Vom Fachbereich Elektrotechnik und Informationstechnik
der Technischen Universität Kaiserslautern
zur Verleihung des akademischen Grades
Doktor der Ingenieurwissenschaften (Dr.-Ing.)
genehmigte Dissertation

von

M. Sc. Yun Wan
geboren in Jiangxi

D 386

Tag der mündlichen Prüfung:	14.08.2017
Dekan des Fachbereichs Elektro- und Informationstechnik:	Prof. Dr.-Ing Ralph Urbansky
Vorsitzender der Prüfungskommision:	Jun.-Prof. Dr.-Ing. Daniel Görge
1. Berichterstatter:	Prof. Dr.-Ing. Steven Liu
2. Berichterstatter:	Prof. Dr.-Ing. Mario Pacas

Acknowledgment

This thesis presents the results of my work at the Institute of Control Systems (Lehrstuhl für Regelungssysteme) in the Department of Electrical and Computer Engineering at the Technical University of Kaiserslautern.

First and foremost, I would like to express my great gratitude to my supervisor Prof. Dr.-Ing. Steven Liu for the excellent supervision of my research work, the valuable scientific discussion and encouragement during my work and research journey at the Institute of Control Systems.

Furthermore, I would like to thank Prof. Dr.-Ing. Mario Pacas for his interest in my research and for joining the thesis committee as a reviewer. My thanks also go to Jun.-Prof. Dr.-Ing. Daniel Görges for joining the thesis committee as a chairman.

My time at the Institute of Control Systems has been very enjoyable and rewarding. I would like to thank all the colleagues for creating an open and cooperative atmosphere. Special thanks go to Priv. Doz. Dr.-Ing. habil. Christian Tuttas, Dr.-Ing. Sanad Al-Areqi, M. Sc. Markus Bell, Dipl.-Ing. Felix Berkel, M. Eng. Sebastian Caba, M. Sc. Filipe Figueiredo, M. Sc. Yanhao He, M. Sc. Hafiz Kashif Iqbal, M. Sc. Jawad Ismail, Dr.-Ing. Fabian Kennel, Dipl.-Ing. Lepold Tobias, M. Sc. Markus Lepper, M. Sc. Guoqiang Li, M. Sc. Xiaohai Lin, Dipl.-Ing. Peter Müller, Dr.-Ing. Philipp Münch, Dr.-Ing. Tim Nagel, Dr.-Ing. Martin Pieschel, Dr.-Ing. Sven Reimann, Dipl.-Ing. Nelia Schneider, Dr.-Ing. Paul Sheuli, Dr.-Ing. Stefan Simon, M. Sc. Tim Steiner, M. Sc. Alen Turnwald, M. Sc. Hengyi Wang, M. Sc. Benjamin Watkins, Dr.-Ing. Jianfei Wang, M. Sc. Min Wu and M. Sc. Yakun Zhou. Thanks also go to the technicians Swen Becker and Thomas Janz and to the secretary Jutta Lenhardt for providing a good technical and administrative environment. Moreover, I would like to thank the visiting professors Prof. Dr. Wenan Zhang and Prof. Dr. Jun Hu as well as the exchange doctoral student Dr. Xusheng Yang for fruitful discussions and valuable suggestions.

I am very thankful to my parents and family for the love and support throughout all the years. Last but never least, I would like to thank my lovely wife Xiyun for standing beside me and inspiring me. This thesis is dedicated to her.

Kaiserslautern, March 2018   *Yun Wan*

Contents

II MMCC direct control method   117

List of Tables

IX

List of Figures

1 Introduction

1.1 Background

Power electronics technology has received great evolution in the recent years due to the new development of power semiconductors, innovations of power converter topologies and control techniques and wide range of applications for everyday uses and industrial production [1]. Generally speaking, power electronics is the technology that realizes the efficient conversion and control of electric energy from the source to the load with the help of switching mode power semiconductor devices, which can be simply classified by its basic conversion tasks into AC/DC, DC/AC, DC/DC and AC/AC. With the trend of information technology, factory automation, home automation and facility management in recent years, an increasing number of power electronic devices are integrated into homes, offices, stores, factories, transportation systems and power systems [2]. The key features of power electronics systems are efficiency, cost reduction and economy, robustness under different conditions, high availability and large production.

The development of power electronics can generally be divided into two trends: to develop novel semiconductors with higher nominal voltages and currents while maintaining the traditional two-level converter structure [3, 4, 5]; and to develop new converter topologies with the existing semiconductors, known as multilevel converters, to satisfy various application scenarios [6]. The research and development of solid-state power semiconductor devices with higher power ratings and improved characteristics concentrates on trials on different semiconductor materials, processing, manufacture and packaging techniques of semiconductors, analysis and modeling of the power switchers and development of intelligent power modules (IPM) [1]. Corresponding to the second trend, new converter topologies are invented and introduced along with improved modeling methods, advanced control and estimator design, new simulation techniques and real-time verification with DSPs and application specified integrated circuit (ASIC) chips. This thesis will introduce a group of emerging multilevel converters with modular construction and cascaded structures, analyze them from modeling and control aspects and summarize a generalized analytical method, which aims to contribute to the second trend.

1

1.2 Multilevel converter technology

Due to the limitation of the voltage ratings of the existing power semiconductors, traditional two-level converter cannot be directly applied to high-voltage large-power applications. Even though the high voltage can be shared by a series stack of semiconductors, it still encounters troubles of equalizing the dynamic and static voltage sharing across each semiconductor, large dv/dt and intensive electromagnetic interference (EMI). Nowadays, multilevel converters have found their significant and wide utilization in medium and high voltage applications, such as, high-voltage direct-current (HVDC) transmission system, flexible alternating current transmission system (FACTS), static synchronous compensator (STATCOM), active power filter (APF), medium-voltage variable-speed drives, railway electric traction system and photovoltaic and wind power system [7]. Compared to two-level voltage source converter, multilevel converters offer the following benefits [8]: (1) Staircase waveform quality. A small portion of injected harmonics is expected and expense for grid-side filter is reduced; (2) A comparatively lower switching frequency. The losses from the power semiconductor is reduced and high-frequency noises is avoid; (3) Reduction of common-mode voltages.

A classification for the multilevel converter topologies is illustrated in Fig. 1.1. Multilevel converters can be classified into diode-clamped converters, flying capacitor (FC) converters and cascaded converters. With respect to the components used for cascaded modules, cascaded converters can be classified as cascaded H-bridge converters with multiple isolated DC sources and the modular multilevel converter (MMC) family with multiple floating capacitors.

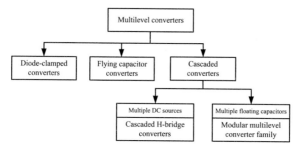

Figure 1.1: Multilevel converter classification

The first multilevel structure began with the three-level diode-clamped inverter, which is proposed in 1981 [9]. The diode-clamped multilevel implementations with different voltage levels are successively reported, and [10] developed a six-level one. The structure of three-phase n-level diode-clamped converter is composed of $n-1$ capacitors, $6(n-1)$ semiconductor switchers and $6(n-2)$ diodes. The $n-1$ capacitors are series connected across the DC link for generating the n voltage levels. The $n-2$ potential nodes between every two neighboring capacitors are attached to the corresponding $2(n-2)$ potential nodes in the upper and lower legs from the same phase with $2(n-2)$ clamped diodes. Principally, an n-level diode-clamped inverter can output n-level phase voltage waveform and $2(n-1)$ level line voltage waveform. However, the diode-clamped PWM converters with a voltage level of more than three encounters great difficulties when putting into market. One reason is that the clamping diodes require higher voltage blocking ability than the free-wheeling diodes across IGBTs, which increases difficulties in converter assembling and reduction of stray inductance. Another reason is that a voltage balancing strategy for the four split DC capacitors can only be achieved with external balancing circuits [11].

[12] introduced a flying-capacitor-based multilevel inverter topology in 1992. The structure of this converter can be directly obtained from the diode-clamped case by replacing the blocking diodes with capacitors. The capacitor voltage difference between the two converter legs generates a voltage level for the output voltage. FC converter has the advantages of redundancies for inner voltage levels, controllability of both active and reactive power and capability of short fault operation and suppress the sudden voltage rises [13]. In addition to the $n-1$ capacitors across the DC link, a n-level FC converter requires $(m-1) \times (m-2)/2$ auxiliary capacitors per phase by selecting the capacitor with the same rating voltage as the nominal voltage of a semiconductor switcher.

Cascaded H-Bridges (CHB) converters, proposed by Robicon Corporation, are capable of generating multilevel voltage waveforms with a separate DC source integrated in each H-Bridge. Each H-Bridge can output the positive and negative DC source voltages as well as the zero voltage by different switching combinations. Therefore, a converter leg with n cascaded H-Bridges can generate $2n+1$ voltage levels. Such implementation can achieve a possible number of voltage levels that is twice of the number of H-Bridges. Since the structure of series connected H-Bridges is suitable for modularized packaging and quick manufacturing, pioneering publications were published focusing on this interesting "modular multilevel" structure. However, a bulky and complicated multi-winding phase-shifted transformer is required to supply electric power for all the floating H-Bridges.

These three types of topologies are effective multilevel converters already validated in laboratory and industrial application. However, they are confronted with their respective limitations and shortcomings. NPC requires a largely increasing number of clamping diodes and the problem of its DC-link voltage balancing becomes severe as the number of voltage levels increases. The same dilemma exists in FC for its massive number of bulky capacitors and their voltage balancing. CHB needs separate DC sources for its H-bridge blocks, which would limit its application to the situation with available multiple separate DC sources. These disadvantages will limit the maximum voltage levels to 5-17 from cost-effective view and make them unsuitable for higher voltage applications [13].

1.3 Novel Modular Multilevel Cascaded Converter (MMCC)

Recently, a group of novel multilevel converters with modular realization and flexible voltage level expansion, termed as Modular Multilevel Cascaded Converter (MMCC), has attracted increasingly concern and popularity in both academic research and industrial application. Compared with the conventional multilevel converters, the MMCC family has become an attractive multilevel solution for medium and high voltage applications [14, 15, 16, 17, 18]. It offers a large amount of voltage levels with linear expanding hardware complexity, less harmonic injection, a modular construction for redundancy, flexible scalability for different voltage requirements and a reduced switching frequency as well as power losses. An MMCC circuit is composed of a cascaded connection of switching submodules (SM) with distinct structural arrangements. It covers a wide range of novel multilevel structures regarding to submodule type, converter branch setup, topological configuration and application scenario.

1.3.1 Modular Multilevel Converter (MMC)

During the last decade, the probably best known MMCC structure, Modular Multilevel Converter (MMC), has found significant industrial utilization for bulk power transmission in VSC-HVDC technology [16, 19]. The MMC contains six converter branches, each of which is constructed by a cascaded connection of a large number of submodules, as shown in Fig. 1.2. The very first publication [20] illustrated his design concept of such modular-constructed multilevel converters for grid connection and summarized the benefits as: 1) redundancy and high availability in electronic component failures, 2) superior harmonics performance with a huge number of voltage levels, 3) a balanced voltage distribution among the power semiconductors, and 4) modularity and scalability to satisfy any voltage level and power requirements.

The MMC was validated under laboratory conditions [21] and Siemens implemented the HVDC PLUS Project with two back-to-back (BTB) MMCs in industrial practice [22] as shown in Fig. 1.3. The milestone of the HVDC PLUS technology is the Trans Bay

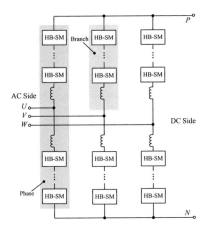

Figure 1.2: MMC with six branches and HB-SMs

Cable Project in Pittburg and San Francisco, which can transmit 400 MW active-power and ±170 Mvar reactive-power along a 85 km submarine cable. The MMC stations in this project contain about 200 submodules per branch, which implement a rectification from 230 kV AC voltages to ±200 kV DC voltage and a inversion from DC to 138 kV AC voltages. Furthermore, MMC is seen as a promising solution for the future high-voltage DC (HVDC) SuperGrids. [18, 23]. As one MMC system is already quite complicated, an HVDC SuperGrid based on multiple MMCs becomes even more intractable. The generalized modeling method discussed in this thesis can standardize and simplify the model deduction of multi-terminal HVDC systems.

1.3.2 Structural development of MMCC

The structural development of MMCC mainly focuses on two aspects: submodule configuration and converter topologies.

Submodule configurations

The most common submodule types are half-bridge (HB-SM) and full-bridge (FB-SM). HB-SM is composed of two semiconductor switchers and one capacitor, which can only generate zero and the positive voltage from the capacitor. FB-SM contains four semiconductors and one capacitor, which can generate zero, positive and negative output voltages. Compared to FB-SM, HB-SM is more advantageous in the AC/DC MMCC

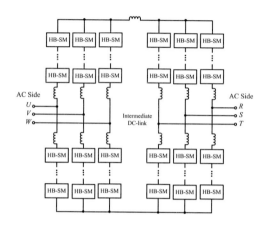

Figure 1.3: Back-to-back MMCs for DC transmission

(MMC) for HVDC application due its lower expenses and reduced power losses. On the other hand, the increased number of semiconductors in HB-SM allows a construction of MMCC that is applicable not only for DC/AC MMCC but also AC/AC MMCC.

Another type of submodule named as double-clamped submodule (DC-SM) was proposed for breaking the high DC current in MMCC-based HVDC application [24]. For a DC-SM, two HB-SMs are series connected together with one additional switcher and two diodes. Such configuration gives the possibility of generating the desired negative voltage for regulating the fault current during a DC-side short circuit. However, it can be found that the maximal possible negative voltage from the capacitors in one branch is only half of the sum of their positive voltage and the power loss for a double-clamped submodule increases by 25 % compared to two HB-SMs due to the conduction loss from the additional semiconductor [25].

Other submodule variations derived from FB-SM were also reported. In [26], a current source submodule was developed by choosing the thyristors with higher voltage and power ratings as switching devices in order to increase the transmitted power. Another variant full-bridge submodule with an additional battery-fed DC/DC converter connected to its capacitor can be adopted in battery energy storage systems (BESS) for decoupling the power variation and capacitor voltage variation from the battery [27].

The traditional multilevel concept can also be introduced to develop the new submodule topologies for the MMCC family. The 3-level neutral-point-clamped submodule and 3-

level flying-capacitor submodule were applied to construct the MMCC topology, which allows a convenient application of the already existing and well-proved multilevel structures under the novel MMCC arrangement [28].

Converter topologies of MMCC

In [29], An HB-SM named as middle cell is added to interconnect the upper branch, the lower branch and the AC terminal point for each phase. The middle-cell-based MMC can achieve the same voltage levels as the traditional MMC with less submodules. However, the redundancy for the middle-cell is difficult to implement so that the reliability of the whole structure is deteriorated. In [25], a so-called Alternate-Arm-based MMC was suggested that is composed of FB-SMs and contains one director switch for each branch. This topology can still retain the ability of controlling the phase currents during the fault due to the director switcher and it reduces the number of the submodules due to application of FB-SMs. Although these alternative MMCs have their improvements, the strict modularity of MMCC structures as well as the reliability are destroyed, which limits their further applications.

[30] introduced MMCC-based static VAR compensator (SVC) with three branches configured in delta connection, which is integrated into the power systems to improve the dynamic stability and power quality [31]. Due to the modularity and scalability to meet any voltage levels, such SVC systems can be directly applied to a medium-voltage grid up to $36\,\mathrm{kV}$ and $\pm100\,\mathrm{MVar}$ without requiring a bulky transformer.

Compared to the indirect AC/AC conversion based on back-to-back MMCs (BTB-MMC) given in Fig. 1.3, a chainlink matrix converter also named as modular multilevel matrix converter (M3C) was developed for implementing a direct AC/AC conversion without the intermediate DC-link [32], which is regarded as a suitable solution for motor drives requiring regenerative braking and overload capability at low speeds [33]. M3C is configured by nine branches with FB-SMs in the 3×3 form of the conventional matrix converter, shown in Fig. 1.4(a). Another interesting direct AC/AC MMCC named as Hexverter was proposed with six branches in a hexagonal configuration [34]. The Hexverter (Fig. 1.4(b)) realizes a direct AC/AC conversion between two 3-phase systems and only requires a smaller number of submodules than BTB-MMC and M3C. It is applied to connect two grids with different frequencies and amplitudes and to drive a motor from a utility grid. However, it was reported that its operation under the equal frequencies of the both systems is greatly constrained and thus not advisable [35].

With all these structural variations and application scenarios in the background, the MMCC technology becomes more and more an important field both for education, research and industry.

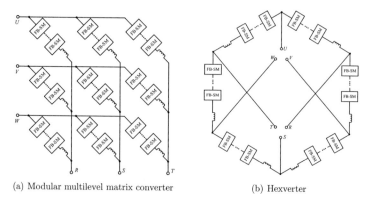

(a) Modular multilevel matrix converter   (b) Hexverter

Figure 1.4: MMCC structures for direct AC/AC conversion

1.4 Motivation

Motivation of MMCC general treatment

The general treatment of MMCC modeling, analysis and control design is, however, not trivial due to the large number of possible topologies, to the different operational requirements for different applications and in many cases simply to the lack of a common understanding of the internal physical relationships. [36] proposes a methodology named Power Block Geometry to depict power electronics topologies, providing a formal and systematic outline for a comprehensive didactic presentation. However, it cannot be employed to modeling the converter dynamics and help to develop controllers. On the other hand, these internal relationships concerning the dynamic states of the high number of submodule capacitors are especially important as they are essential for achieving high-quality control. In most cases, however, modeling of an MMCC has been carried out for a concrete configuration and a specific operation as well as control goals. By choosing case-specific system variables and formulating their equations, it usually leads to a case-specific MMCC model. Although these models are usually efficient for their special purposes, it often turns out that they do not provide a common physical interpretation, which makes the portability of the developed analytical methods to other MMCC topologies a difficult issue. Furthermore, it is not an easy task to present the unified method to generalize the MMCC family especially for the educational purpose.

In addition, considering the common characteristics of the MMCC family, the continuous branch current even with switched branch submodules and the distributed energy storage units, it makes a unified modeling and analysis not only reasonable, but also possible. Generally speaking, an MMCC is an energy-storage system that serves as the link between up to two external voltage systems. Thus, to realize the external functionality and to guarantee internal distributed energy balancing are the two main tasks of MMCC control. The external functionality of an MMCC as well as the regulation of the total stored energy is fulfilled by its external currents relative to the generated equivalent external voltages from MMCC branch voltages. The additional degrees of freedom provided by the branch voltages allow introducing an internal-flowing current, named circulating current (CC), to adjust the internal energy distribution among these branches. The analysis and control of the CCs can be organized, in spite of the topological differences, in a unified way by manipulating them in specific shapes to realize the internal branch energy balancing (IBEB). The state of IBEB guarantees a safe long-term operation of an MMCC-based system. Finally, a underlying energy balancing to the level of single submodule capacitor is achieved by the proposed IBEB strategy combined with MMCC modulation techniques based on submodule voltage sorting. In this thesis, these physical principles will be analyzed and a general methodology with two layers (current and energy layers) and two scopes (external and internal scopes) is developed to model and analyze the MMCC family.

Besides the technical issues mentioned above, another essential aspect for developing a unified framework is a common describing and naming tool to represent the large number of MMCC topologies. For this purpose an efficient topological describing method will be established based on graph theory prior to the modeling procedure, giving the systematic treatment a formal basis.

Motivation of direct control

An important issue regarding MMCC is to control its multiple independent current and balance distributed submodule voltages through selection among numerous switching combinations in every control period. The classic control structure for an MMCC system is composed of a multivariable controller for its current and energy variables and a multilevel modulation block embedded with submodule-voltage-sorting algorithm. The modulation block basically implements the task of "translating" the continuous branch voltage commands into the discrete ON-OFF signals for the branch submodules, and meanwhile balances the voltage distribution among the submodules. The commonly-used MMCC modulation method adopts the assumption that all the submodules have the same capacitor voltages equal to the constant reference voltage, which totally applies for high-voltage application with hundreds of submodules per branch. For MMCC-based medium-voltage application that the number of branch submodules is low and the switching frequency is restricted, it is almost impossible to ensure this assumption due to the following two facts:

1. All the submodules have voltage ripples around the reference voltage during operation. Since the branch current actually flows through the switched-on submodules,
   the voltage ripples of submodule capacitors can never be eliminated. It is possible to suppress these ripples by increasing submodule capacitance, the number of
   branch submodules or line frequency, which tremendously raises the system cost.

2. During MMCC operation the instantaneous submodule voltages vary in time and
   differ with each other. Additionally, the inhomogeneous distribution of submodule parameters, such as submodule capacitance and turn-on resistance, leads to
   distinct submodule charging and discharging dynamics.

The above two known facts will increase the difficulty of generating the appropriate
switching combination to guarantee the quality of MMCC external currents and balancing status of submodule voltages. It becomes particularly critical for a medium-voltage
application when the number of branch submodules is low.

In order to cope with the above dilemmas, the concept of direct torque control (DTC) for
motor drive is deployed to develop direct control strategy for MMCC, which generates
the switching states for branch submodules according to requirement of external current
control and internal voltage balancing and removes the intermediate "translation" stage.
In comparison with the conventional direct torque control based on two-level inverter,
an increasing number of switching combinations are required to be evaluated for MMCC
direct control, which makes the design and solution of direct control method an intractable issue. In the second part of this thesis, direct control method will be developed
based on MMCC hybrid switched model and directly determines the optimal switching
states for all the submodules by minimizing a predefined integrated current-energy cost
function. MMCC direct control aims to establish the explicit relationship between submodule switching states and the comprehensive control objectives, e.g., external current
control, internal energy balancing and submodule voltage ripple regulation, with respect to unequal instantaneous submodule voltages and inherent submodule parametric
inhomogeneity.

1.5 Objectives and outlines

1.5.1 Objectives

Despite numerous publication on the subject of MMCCs, the corresponding modeling
and control issues are handled from the perspectives isolating from the MMCC family.
Thus, it results in a research status that MMCC analysis is a collection of separated

reports on one particular MMCC structure with different technical methods, which increases the difficulty of understand this entire novel converter family. This thesis aims to establish a physical interpretation of MMCC and develop two frameworks for the control-oriented MMCC modeling as well as the hierarchical analysis. One framework analyzes MMCCs as continuous systems on the basis of treating each converter branch as controllable voltage and controllable current sources. Another one is to maintain the switching features of MMCC switches and handle the problems in the context of hybrid systems. Accordingly, the entire thesis is divided into two parts.

Objectives of Part I

Part I presents an analytical framework that enables a unified approach of modeling and analysis for the MMCC family. In comparison to the conventional method of analyzing each MMCC structure based on special applications, the presented methodology proposes a generalized viewpoint for dealing with all kinds of MMCCs. The developed integrated current-energy model from this analytical framework has a standard state-space formulation containing the system features. After that, a collective group of control methods based on multivariable optimal control law are accordingly applied to regulate the large quantity of system states inside MMCC and achieve their respective optimality. The involved contribution is listed as follow.

1. MMCC topological analysis based on graph theory is presented. This framework allows a structural grouping of MMCC branches and provides the basis of a formal description of the proposed modeling and analysis.

2. The proposed generalized modeling method divides MMCC modeling into four decoupled parts, which are external current modeling, external power modeling, internal circulating current modeling and internal branch energy balancing. By strictly following this method, a standard state-space model of an MMCC system can be obtained that embodies its model features of multi-input and multi-output, dynamical coupling effects, nonlinearity and time-variance.

3. To demonstrate the suitableness of the developed framework for the control purpose, a generalized control design combining multivariable control theory and optimal principle is proposed. It is especially suitable for MMCC system based on multivariable models, which systematically handles the control issues of MMCC external functionality and internal energy balancing.

Objectives of Part II

Part II aims to contribute modeling and control of MMCC system from the perspective of switched submodules, which evaluates the influence of submodule switching to MMCC system layer as well as the branch layer. Taking the MMCC-based medium-voltage application into account, the conventional intermediate modulator that introduces time delay and modulation error is eliminated in direct control method. A detailed list of the contribution is given here.

1. A direct switched model of MMCC is developed to establish an explicit relation between submodule switching states and system state variables, e.g., independent current variables, branch energies and submodule voltages. It is based on the analytical framework in Part I and extends to the very bottom layer of submodule control. Based on the direct switched model, problem formulation of MMCC direct control is presented, which is a multi-step discrete optimization problem.

2. A practical solution method based on delta-input formulation and control set reduction is presented. It greatly reduces the computation effort for balancing the submodule voltages compared with complete enumeration and reserves dynamic performance for external current tracking.

3. Event-based direct control method employs a distributed computing architecture that assigns one event generator to each submodule, which produces a switching index with respect to current submodule state. In each control period a dynamic control set is constructed by referring to the switching indices, and it is evaluated for obtaining the optimal switching combination. The event-based method is characterized by adaptively long prediction horizons, rationally reduced switching frequency as well as the lowest time complexity.

1.5.2 Outlines

Part I (from Chapter 2 to 6) focuses on the unified framework of modeling and analysis for MMCC, and Part II (from Chapter 7 to 9) is devoted to the framework of MMCC direct control.

Chapter 1 gives an overview of multilevel technology and the emerging MMCC systems. The motivations of MMCC general treatment and direct control are explained, and the contribution of this thesis is clarified.

Chapter 2 begins with an explanation of the basic concepts of MMCCs that are frequently used in this thesis, which are submodule, branch, topology, structure and MMCC system. Then two classification methods based on energy conversion types and based on 3-phase connections are presented. In Section 2.3, a comprehensive classification based on graph theory is proposed so that MMCCs can be classified into bipartite-graph-based and cycle-graph-based types. Additionally, branch grouping for MMCC branches is explained, which offers a structural division and simplification for MMCC topologies.

Chapter 3 focuses on the unified framework of modeling and analyzing MMCCs as continuous systems. Section 3.1 proposes a continuous equivalent modeling method for analyze branches and MMCC systems. The two decoupled modeling layers for current dynamics and the energy dynamics are explained in Section 3.2 and 3.3, and finally integrated in Section 3.4. In each layer two scopes for external and internal dynamics are explained, constituting the entire behavior description.

Chapter 4 begins with modern multivariable control design based on the state-space model in Section 4.1. An efficient optimal control design, Linear Quadratic Regulator (LQR), is introduced in Section 4.2. Section 4.3 handles the nonlinearity in the integrated MMCC model by applying nonlinear quadratic control based on the state-dependent Riccati equation. Section 4.4 defines the hyper period for typical MMCC systems and presents periodic discrete LQR to deal with time-varying feature. Section 4.5 discusses the modulation techniques applied for MMCC.

Chapter 5 gives three case studies by following the modeling method, which are MMC-based medium-voltage direct current transmission system for active power transmission, delta-configured MMCC for reactive power compensation and modular multilevel matrix converter for connecting two three-phase AC grids.

Chapter 6 describes the implemented three-phase MMC system in laboratory and presents the according experimental results.

Chapter 7 starts to focus on direct control method. A complete procedure of developing direct switched model that describes the explicit relationship between submodule switching combination and system states is given by taking one-phase MMC as example. Based on the model, direct control problem is formulated as multi-step discrete optimization with an integrated cost function and a huge total control set.

Chapter 8 firstly explains the tasks of specification of MMCC direct control. In Section 8.2 two enumeration algorithms that determine the optimal solution of discrete optimization problem are discussed, which are exhaustive search and dynamic programming. Section 8.3 presents fast reduced control set method that reduce the total control set to a reduced size by introducing physical consideration of avoiding unnecessary switching, referring to switching history and predicting the future states. In Section 8.4 the proposed event-based method assigns one event generator to each submodule. The event generator evaluates submodule switching capacity based on current states, predicted states and individual parameters and assists the central optimizer to find the most suitable switching decision.

Chapter 9 presents the simulation study and experimental validation of the one-phase MMC under direct control method. The performance of direct control is evaluated from switching frequency, voltage&current THD, submodule voltage stability and time complexity by taking the classic voltage-sorting method as a benchmark. Section 9.2 shows the experimental results of direct control method on a laboratory MMC prototype.

Chapter 10 summarizes the achieved work of this thesis.

Part I

Unified framework of modeling and analysis for MMCC

2 Topological analysis of MMCC

In order to clearly distinguish the diverse MMCCs and precisely address them, a terminology system is demanded. The existing terminology methodologies categorize MMCC structures either with energy conversion types or 3-phase connection types, which do not provide a comprehensive understanding of the whole family and a precise way for the following mathematical modeling. In this chapter, a graph-based analytical method is proposed to abstract the MMCC topologies and develop a systematic classification and terminology methodology. Compared to the existing classification methods, the proposed methodology precisely classifies all the existing MMCC structures based on their graph types and submodule types. It also allows a systematic way of assigning specific symbols for topology names, terminals and branches, which makes the further description for the generalized modeling and analytical method in Chapter 3 clear and rational.

In Section 2.1, the basic concepts of MMCCs frequently quoted in this thesis are firstly described, which are submodule, branch, MMCC topology, MMCC structure and MMCC system. Section 2.2 discusses the two classic classification methods and analyzes their respective advantages and limitations. In Section 2.3, the graph information of all the MMCC topologies is extracted and used as a reference for the improved graph-based terminology methodology. With the help of such formal description system based on graph theory, a systematic terminology from MMCC terminals to its branches until the overall topology is achieved.

2.1 Basic concepts of MMCC

To understand MMCC, submodule and branch for MMCC should firstly be introduced. Submodule is usually composed of a fixed number of semiconductor switches allowing a generation of certain voltage levels at the submodule terminals. An arbitrary number of identical submodules are connected in series and construct an MMCC branch with a branch inductor. A fixed number of branches are configured according to certain topological types to form an MMCC structure. The MMCC structure cannot implement its functionality unless it is interconnected with other external systems (voltage sources, motors and other power loads) to constitute an MMCC system. Therefore, its structural order is listed as semiconductor switch, submodule, branch, MMCC structure and finally MMCC system, also providing a layered concept of analyzing MMCCs. First of all, submodule in MMCC is presented.

2.1.1 MMCC fundamental cell - Submodule

The identical submodules (SM) are connected in series, so that they are scalable to different power- and voltage ratings and expandable to diverse voltage levels. Fig. 2.1(a) and 2.1(b) show the two commonly-used submodule configurations: the half-bridge submodule (HB-SM) and the full-bridge submodule (FB-SM). Some other topologies can also be referred to develop MMCC submodules, such as, double-clamped submodule shown in Fig. 2.1(c), cascade-half-bridge submodule and 3-level neutral-point-clamped submodule.

Half-bridge submodule

A HB-SM contains two IGBTs with two anti-parallel diodes and one capacitor, which can be operated in three different states: ON, OFF and Block. As shown in Fig. 2.1(a), ON-State corresponding with IGBT1 switched on and IGBT2 switched off outputs the positive voltage U_c across the capacitor, whereas OFF-State switching IGBT1 off and IGBT2 on leads to a short cut and outputs zero voltage. For Block-State, both IGBTs are switched off, which is helpful to precharge the submodule capacitor and prevent submodule damage in fault conditions.

Full-bridge submodule

A FB-SM is composed of four IGBTs with four ant-parallel diodes and one capacitor, which constitutes a single-phase full-bridge, or H-bridge. Compared to HB-SM, FB-SM can output not only zero and positive capacitor voltage $+U_c$, but also negative capacitor voltage $-U_c$ by adjusting the gate signals for IGBTs. Therefore, the switching states for a FB-SM can be divided into: Positive, Zero and Negative. When IGBT1 and 4 are switched on while IGBT2 and 3 off, a conductive path can be constituted at FB-SM's terminals to output the positive capacitor voltage U_c. On the contrary, when switching only IGBT2 and IGBT3 on, a conductive path for outputting the negative capacitor voltage $-U_c$ is formed at FB-SM's terminals. A short circuit at its terminals can be obtained either by switching the both upper IGBTs on or by switching the both lower IGBTs on.

Double-clamped submodule

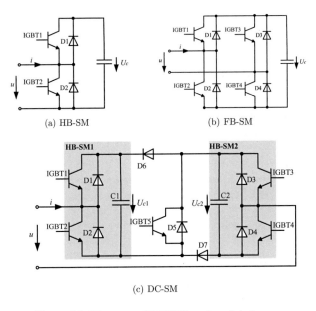

(a) HB-SM   (b) FB-SM

(c) DC-SM

Figure 2.1: Diagrams of MMCC's submodule types

An HB-SM-based MMC is vulnerable to a DC-side fault, since a reliable high-voltage DC breaker is still quite challenging and such MMC cannot produce a negative branch voltage to suppress the generated DC fault current. Although the FB-SM can inherently produce a negative voltage, the expense associated with the component cost and power loss of FB-SM-based MMC are double of HB-SM-based one. Therefore, the double-clamped submodule (DC-SM) was then suggested as an effective and economical submodule type to solve this problem [24]. A DC-SM is composed of two HB-SMs (HB-SM1 and HB-SM2) and an additional power switch (IGBT5) as shown in Fig. 2.1(c). For a normal operation, IGBT5 is assigned to be switched on and constitutes a bidirectional current path with the anti-paralleled diode D5. When detecting a DC-side fault current, all the five IGBTs are switched off and the paths to suppress the fault current can be constituted with respect to different directions of i.

- If $i > 0$, the fault current i flows through D1, C1, D5, C2 and D4. A voltage $-U_{c1} - U_{c2}$ is then generated to break the high fault current i.

- If $i < 0$, diodes D3, D6, D7 and D2 conducts and the two capacitors (C1 and C2) are parallel connected to share the current i. That is to say, only U_{c1} (assume

$U_{c1} \approx U_{c2}$) can be produced to resist negative current flow i.

From the above analysis, it can be seen that the ability of breaking the fault DC current reduces by half if $i < 0$. As a conclusion, by introducing one additional IGBT for each two HB-SMs, the obtained DC-SM permits MMC to break fault DC currents.

Cascade-half-bridge submodule

By replacing each IGBT with a cascaded connection of IGBTs in an HB-SM, the cascade-HB-SM can be developed [37]. Compared to the HB-SM in Fig. 2.1(a), the cascade-HB-SM can inherently stand a higher voltage and the required submodule number for a fixed voltage rating can also be reduced. It is clear that a reduction of branch submodules can simplify the control issue and potentially improve the system reliability, although the number of the resulting voltage levels is also reduced. Therefore, a compromise between the number of branch submodules and the number of cascaded IGBTs can be achieved. However, it is quite challenging to achieve a voltage balancing among the series-connected IGBTs due to parameter differences of the semiconductor switches.

Multilevel submodule

The multilevel submodule is developed by employing the conventional multilevel concepts that already exists for a long time and has been examined [28]. A higher rating voltage for each submodule associated with a reduction of branch submodules can be realized for a given voltage rating without reducing the number of voltage levels. However, it must be pointed out that the modulation method and capacitor voltage balancing for the branch submodules could be highly complicated and should be carefully chosen with respect to different multilevel submodules.

All these submodule types can be classified into unipolar, bipolar and multipolar submodules, which can be seen in Table 2.1. Suppose the submodule capacitor has a voltage of U_c.

2.1.2 Basic component - Branch

An MMCC branch is composed of N cascaded-connected identical SMs and a series branch inductor, as shown in Fig. 2.2. The N submodules in the same branch is called branch submodules. Only the voltage across the branch submodules is defined as the branch voltage. In other words, the branch voltage does not contain the voltage drop from the branch inductor. The functional benefits of branch inductor are as follows. It can suppress and control the possible circulating currents in MMCCs that are excited by the current loop constituted by the branches. Additionally, the branch inductor can

Table 2.1: Submodule types for MMCC structures

Basic Type	Subtype	Output voltage	Evaluation
Unipolar SM	HB-SM	$0, U_c$	Typical unipolar SM
	Cascade-HB-SM	$0, U_c$	Increase voltage rating for a single SM
Bipolar SM	FB-SM	$-U_c, 0, U_c$	Typical bipolar SM
	DC-SM	$0, U_c, 2U_c$. In fault: $-2U_c, U_c$	Allow breaking fault DC current by adding one switch for every two SMs
Multipolar SM	NPC-SM	Multilevel output voltages	Increase voltage rating for a single SM without reducing the voltage levels
	FC-SM		
	NPP-SM		

reduce the sudden rising currents during the faults occurring inside or outside an MMCC. The step-wised branch voltage appears at MMCC terminals smoother and contains less harmonics due to the branch inductor.

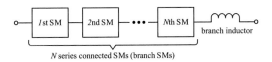

Figure 2.2: Diagram of MMCC branch

Combined the discussion of the discrete output voltage values of submodules in Table 2.1, the possible discrete branch voltage across the cascaded SMs can also be analyzed. For example, N HB-SMs are connected in series and each HB-SM is assumed with an equal capacitor voltage U_c, the branch voltage can be chosen from the values given in the set below:

$$\mathcal{U}_{HB-br} = \{\ 0,\ U_c,\ ...,\ (N-1)U_c,\ NU_c\ \}$$

In the case of a branch with N FB-SMs, its branch voltage is limited in the following set

$$\mathcal{U}_{FB-br} = \{\ -NU_c,\ -(N-1)U_c, ...,\ -U_c,\ 0,\ U_c,\ ...,\ (N-1)U_c,\ NU_c\ \}$$

The possible voltage levels of a branch with other submodule types can be analyzed similarly, which is neglected here. Without loss of generality, only the two most common submodule types (FB-SM and HB-SM) are focused in the later classification.

2.1.3 MMCC topology, structure and system

In this subsection, three terms "MMCC topology", "MMCC structure" and "MMCC system" are explained, which will be frequently quoted later.

MMCC topology

The noun "circuit topology" has been commonly-used in explaining power electronics devices that describes the main configuration of voltage sources, inductors, capacitors, semiconductors and loads for a power converter. For example, the difference between buck topology and boost topology focuses on the relative placement of the involved passive components and the solid state switches. However, for describing the complicated MMCC containing hundreds and even thousands of switches and a number of branch inductors, it is not suitable to define "MMCC topology" at the component level due to overcomplexity of description and incapability of capturing the feature. Thus, branches are then selected as the basic unit to feature an MMCC regardless the exact branch configuration. In this thesis, MMCC topology is defined to represent the branch connectivity inside an MMCC.

MMCC structure

The term "MMCC structure" is assigned to describe the exact schematic of an MMCC, which considers not only branch connectivity but also the exact branch configuration and the submodule type. Therefore, the same MMCC topology can evolve to several MMCC structures after giving some adaption to the branch construction and adopting different submodule types. For example, MMC with six branches and HB-SMs shown in Fig. 1.2, the MMC with FB-SMs instead of HB-SMs, middle-cell-based MMC and alternate-arm-based MMC are four different MMCC structures, but they belong to the same MMCC topology.

MMCC system

An MMCC structure cannot implement its functionality without connecting to external systems. These external systems include different types of voltage sources even power grids, electrical machines and other power loads. The "MMCC system" means an MMCC structure with its external connected systems. For example, when one MMCC connects an AC voltage source and an electric motor, the MMCC system means the integration of MMCC, the AC sources and the motor. Another example is that one three-terminal MMCC is connected to three-phase AC grid for reactive power compensation. This MMCC, the AC grid and the possible load constitute the MMCC system. Therefore, it is necessary to analyze MMCC operation in the context of its connected external systems and to implement modeling and control for the investigated MMCC system.

2.2 Classic MMCC classification methods

Since the MMCC family contains a large number of MMCC structures with various topological configurations and diverse submodule types, it is necessary to develop a classification and terminology method so that each MMCC structure can be clearly distinguished as well as precisely addressed. Two classic classification methods are discussed in this section, which are based on energy conversion types proposed by Prof. Marquardt and based 3-phase connections proposed by Prof. Akagi.

2.2.1 Classification based on energy conversion types and 3-phase connections

As power converters are developed to realize different energy conversion tasks, MMCCs can thus be classified according to the energy conversion types [21]. The two basic types of MMCCs are DC/AC and AC/AC conversions. Each basic type can be further divided into 1-phase, 3-phase and multi-phase subtypes regarding their AC-side phases. Table 2.2 summarizes the classification results.

Table 2.2: MMCC Classification based on energy conversion types

Basic type	MMCC subtype	Topology name	Submodule type
DC/AC	DC/1-phase	Single-phase MMCC [38]	HB-SM
	DC/3-phase	MMC	HB-SM
	DC/multi-phase	Multi-phase MMCC	HB-SM
AC/AC	1-phase/1-phase	Single-phase MMCC	FB-SM
	1-phase/3-phase	MMC	FB-SM
	3-phase/3-phase	3-phase/3-phase MMCC (M3C) [32]	FB-SM
	multi-phase ACs	Multi-phase MMCC	FB-SM

The classification method firstly proposed in [15] and later supplemented with Hexverter and M3C in [39] classifies the MMCC family according to the arrangement types of the branches into star-connection- and delta-connection-based MMCCs. Single-star full-bridge (SSFB)- and Single-delta full-bridge (SDFB)-MMCC are the two primary circuit configurations, which are "doubled" and "tripled" into the new ones. The classification based on 3-phase connection is given in Table 2.3.

Table 2.3: MMCC classification based on 3-phase connections

3-phase type	Configuration	Topology	Submodule type
Star-connection -based MMCCs	Single-star	SSFB-MMCC	FB-SM
	Double-star	MMC	HB-SM
		DSFB-MMCC	FB-SM
	Triple-star	M3C	FB-SM
Delta-connection -based MMCCs	Single-delta	SDFB-MMCC	FB-SM
	Double-delta	Hexverter	FB-SM

2.2.2 Comments on both methods

The classification method based on energy conversion types, short as ECT-based classification, focuses on the functionality of MMCC, and meanwhile the classification method based on 3-phase connections, short as 3PC-based classification, focuses on the structure (topology configuration and submodule type) of MMCC. The comparison is summarized in Table 2.4.

Table 2.4: Comparison between two classification methods

	ECT-based Classification	3PC-based Classification
Completeness	Comparatively complete	Limited to three-phase MMCCs
Not included	SSFB, SDFB, Hexverter	Single- and multi-phase MMCCs
Accuracy	Some confusion	Unique terminology
Terminology rule	Universally applicable, but not precise	Simple and precise.

ECT-based classification is a function-oriented method, which is easily extended to the emerging MMCC structures, such as unclassified SSFB-, SDFB- and DDFB-MMCCs. From this sense, 3PC-based classification is less general, as single-phase and multi-phase MMCs cannot be categorized into its system. However, structure-oriented classification extracts the structural features of MMCCs and provides an exact image for the listeners. It allows an MMCC structure to be uniquely determined by its topology type and its submodule type. Thus, 3PC-based classification indicates clearly the concrete MMCC structure and accordingly provide a precise terminology for MMCCs. As the both methods have their own limitations, the graph-based classification will be presented in the following section, which provides a comprehensive classification and an accurate terminology for the MMCC family.

2.3 Graph-based analysis - A formal description

Through a reference with graph analysis, a graph can be associated with an MMCC topology, whereby vertices and edges of a graph correspond to terminals and branches of an MMCC topology. As the first step a formal description based on graph theory is proposed for MMCCs that allows systematic terminology from terminals to branches until topology types.

2.3.1 MMCC description based on graph theory

An MMCC system is generally depicted in Fig. 2.3, which is composed of the MMCC and two external systems, System1 and System2. The connected external systems represent the physical electrical connections at the MMCC structure. External current flows are assumed between the external system and the MMCC, which realizes the requested power exchange. All the MMCC branches carry continuous branch currents during operation, meaning that the branch configuration maintains unchanged and no commutation exists among MMCC branches. Although the dynamic commutation is still assumed in each submodule, MMCC can be treated as a continuous system at the branch layer while its branch connectivity determines the external current distribution. It distinguishes MMCC from conventional converters featured as hybrid switched system.

By introducing the graph theory to analyze the MMCC family, a correspondence between an MMCC structure and a graph can be established, whereby the terminals and branches of an MMCC correspond to the vertices and edges of a graph. In Fig. 2.3 the M terminals at System1 side are named as $\{v_{1,j}\}$ ($j = 1, ..., M$) while the N terminals at System2 are denoted as $\{v_{2,k}\}$ ($k = 1, ..., N$). All the $N_{tm} = M + N$ terminals form the vertex set noted as $\mathcal{V} = \{\{v_{1,j}\}, \{v_{2,k}\}\}$. Between the terminals from System1 and the ones from System2, there exists a fixed number of branches. These branches can be uniquely replaced by lines connecting between the vertices, and these lines are the edges from the point of view of graph theory. An arbitrary branch that connects Terminal $v_{1,j}$ to Terminal $v_{2,k}$ can be denoted as Edge $E_{j,k}$, or alternatively written as $v_{1,j}v_{2,k}$. All the N_{br} edges constitute the edge set $\mathcal{E}$, where $\mathcal{E} = \{E_{j,k}\}$. Additionally, the positive directions for the branch current $i_{j,k}$ and the branch voltage $u_{j,k}$ in Branch $E_{j,k}$ are determined as the directional flow from Terminal $v_{1,j}$ to $v_{2,k}$.

A graph containing the vertex set $\mathcal{V}$ and the edge set $\mathcal{E}$ is denoted as $\mathcal{X} = \{\mathcal{V}, \mathcal{E}\}$,

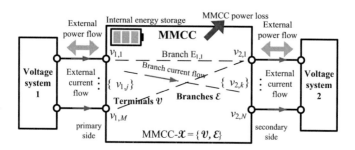

Figure 2.3: An illustrative diagram for an MMCC and its connections

which can also present an MMCC topology. Additionally, branch connectivity of an MMCC topology is described by the graph adjacent matrix $M_{aj}(\mathcal{X})$, which is helpful to identify its graph type. Table 2.5 summarizes the consistent one-to-one matching between MMCC topology and graph. According to graph theory, MMCC graphs can be classified into bipartite graph and cycle graph.

Table 2.5: Correspondence between MMCC topology and graph

Topology MMCC-$\mathcal{X}$	Graph $\mathcal{X} = (\mathcal{V}, \mathcal{E})$
N_{tm} terminals	Vertices $\mathcal{V} = \{\{v_{1,j}\}, \{v_{2,k}\}\}$
N_{br} branches	Edges $\mathcal{E} = \{E_{j,k}\}$
Branch connectivity	Adjacent matrix $M_a(\mathcal{X})$
Branch current $i_{j,k}$	Directional flow from $v_{1,k}$ to $v_{2,k}$
Branch voltage $v_{j,k}$	Directional flow from $v_{1,k}$ to $v_{2,k}$

2.3.2 Bipartite-graph-based MMCCs (Group $\mathcal{K}_{M,N}$)

A complete bipartite graph with one partition containing M vertices and another partition containing N vertices is denoted as $\mathcal{K}_{M,N}$. Strictly speaking, the classified bipartite-graph-based MMCCs, symbolized as MMCC-$\mathcal{K}_{M,N}$, belong to complete bipartite graphs. For MMCC-$\mathcal{K}_{M,N}$, its vertex set $\mathcal{V}$ admits a partition into 2 classes (V_1 and V_2), which satisfies:

$$\mathcal{V} = V_1 \cup V_2 \tag{2.1}$$
$$V_1 = \{v_{1,j}\} \quad j = 1,\ 2,\ ...,\ M \tag{2.2}$$

$$V_2 = \{v_{2,k}\} \quad k = 0, \ 1, \ ..., \ N \tag{2.3}$$

The edge set $\mathcal{E}$ of an MMCC-$\mathcal{K}_{M,N}$ is given as

$$\mathcal{E} = \{E_{j,k}\} = \{v_{1,j}v_{2,k}\} \tag{2.4}$$

The graphs for bipartite-graph-based MMCCs can be plotted in Fig. 2.4, which are marked as $\mathcal{K}_{2,1}$, $\mathcal{K}_{2,2}$, $\mathcal{K}_{3,1}$, $\mathcal{K}_{3,2}$ and $\mathcal{K}_{3,3}$ in sequence. The general case of a multi-phase MMCC is not plotted here. A brief explanation is given to these graph symbols in order that the readers can relate to the MMCC topologies. The MMCC-$\mathcal{K}_{2,1}$ is usually applied with HB-SMs to verify the one-phase operation of the classic MMC [40, 41, 39]. MMCC-$\mathcal{K}_{2,2}$ is actually the single-phase MMC. MMCC-$\mathcal{K}_{3,1}$ constructed by FB-SMs is single-star full-bridge (SSFB)-MMCC, MMCC-$\mathcal{K}_{3,2}$ is the topology derived from the classic MMC, and $\mathcal{K}_{3,3}$ represents the topology of a modular multilevel matrix converter.

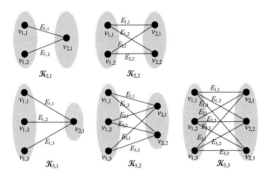

Figure 2.4: Bipartite-graph-based MMCCs

2.3.3 Cycle-graph-based MMCCs (Group $\mathcal{C}_{M+N}$)

A cycle graph is a graph that consists of a single cycle. A cycle graph can be easily understood from a graphical sense, even though a strict definition of cycle graph is based on definitions of path and cycle.

The graphs for cycle-graph-based MMCCs are plotted in Fig. 2.5 and marked as $\mathcal{C}_3$ and $\mathcal{C}_6$. For a better understanding, the positions of certain vertices are adjusted to form the cycle. The cycle-graph-based MMCCs are represented by the symbol $\mathcal{C}_{M+N}$ where M and N are the terminal numbers of the both-side systems.

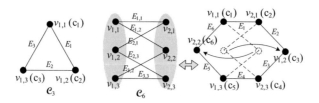

Figure 2.5: Cycle-graph-based MMCCs

For the cycle graph $\mathcal{C}_3$ ($M = 3$, $N = 0$), it satisfies

$$\mathcal{C}_3 = (V_1, \mathcal{E}) \tag{2.5}$$
$$V_1 = \{v_{1,1}, v_{1,2}, v_{1,3}\} \tag{2.6}$$
$$\mathcal{E} = \{E_1, E_2, E_3\} \tag{2.7}$$

Based on the cycle information, the branches are termed in a clockwise direction as E_1, E_2 and E_3, and so the terminals can also be termed as c_1, c_2 and c_3. MMCC-$\mathcal{C}_3$ is the topology of single-delta full-bridge (SDFB) MMCC.

For $\mathcal{C}_6$ ($M = 3$, $N = 3$), it is expressed as

$$\mathcal{C}_6 = (V_1 \cup V_2, \mathcal{E}) \tag{2.8}$$
$$V_1 = \{v_{1,1}, v_{1,2}, v_{1,3}\}, \quad V_2 = \{v_{2,1}, v_{2,2}, v_{2,3}\} \tag{2.9}$$
$$\mathcal{E} = \{E_{1,1}, E_{1,2}, E_{2,1}, E_{2,3}, E_{3,2}, E_{3,3}\} \tag{2.10}$$

After adjusting the vertex positions, it can be seen that $\mathcal{C}_6$ represents the Hexverter topology. For the MMCCs from Group $\mathcal{C}_{M+N}$, all the vertices have an exact degree of 2, which leads to determine a cycle-graph-based MMCC.

As a conclusion, the MMCC family can be classified and symbolized based on the graph types and submodule types as shown in Table 2.6. For simplification, a typical unipolar submodule, HB-SM, and a typical bipolar submodule, FB-SM, are used to illustrate their structures.

2.3.4 Evaluation of graph-based classification

This classification method achieves the following benefits. Firstly, all the existing MMCC structures can be categorized under their graph types and submodule types. Secondly, each MMCC structure can be uniquely identified after this classification. For example, the single-phase MMC applied to implement a direct 1-phase/3-phase conversion

Table 2.6: Graph-based MMCC classification

Graph type	Subtype	SM type	Functionality	Terminology
Bipartite graph	$\mathcal{K}_{2,1}$	HB	DC/1-phase	MMCC-HB-$\mathcal{K}_{2,1}$
	$\mathcal{K}_{2,2}$	FB	1-phase/1-phase	MMCC-FB-$\mathcal{K}_{2,2}$
	$\mathcal{K}_{3,1}$	FB	STATCOM	MMCC-FB-$\mathcal{K}_{3,1}$
	$\mathcal{K}_{3,2}$	HB	DC/3-phase	MMCC-HB-$\mathcal{K}_{3,2}$ (MMC)
		FB	1-phase/3-phase	MMCC-FB-$\mathcal{K}_{3,2}$
	$\mathcal{K}_{3,3}$	FB	3-phase/3-phase	MMCC-FB-$\mathcal{K}_{3,3}$ (M3C)
	$\mathcal{K}_{M,N}$	FB	Multi-phase ACs	MMCC-FB-$\mathcal{K}_{M,N}$
Cycle graph	$\mathcal{C}_3$	FB	STATCOM	MMCC-FB-$\mathcal{C}_3$
	$\mathcal{C}_6$	FB	3-phase/3-phase	MMCC-FB-$\mathcal{C}_6$ (Hexverter)

is termed as MMCC-FB-$\mathcal{K}_{3,2}$, while the classic MMC is termed as MMCC-HB-$\mathcal{K}_{3,2}$. Thirdly, all the terminals and branches are uniformly symbolized, which helps the later mathematical description of the generalized modeling method. In the end, a structural simplification, namely branch grouping, can be developed based on the adjacent branches at a vertex. Branch grouping is helpful for obtaining dq models of an MMCC system and also for developing the internal branch energy balancing (IBEB) model. Branch grouping will be discussed in the next subsection.

2.3.5 Branch grouping

The branch grouping is to group the MMCC branches containing the same terminal, which simplifies the MMCC model, enables a feasible $\alpha\beta$ and dq transformation and facilitates the MMCC branch energy balancing. From the point of view of graph theory, branch grouping is to categorize all the adjacent edges at each vertex in a partition. The definitions of branch group and branch grouping are given here.

Definition 2.1. *A collection of all the branches which share the same terminal $v_{1,j}$ or $v_{2,k}$ is called the branch group with respect to the sharing terminal $v_{1,j}$ or $v_{2,k}$, symbolized as $BG(v_{1,j})$ or $BG(v_{2,k})$.*

Definition 2.2. *Branch grouping for an MMCC $\mathcal{X}$ is the act of classifying all the branches of the MMCC into a certain number of branch groups with respect to all the terminals belonging to the same external system (System1 or System2), which is denoted as $\mathcal{P}(\mathcal{X}, S1)$ or $\mathcal{P}(\mathcal{X}, S2)$.*

Example For M3C (MMCC-FB-$\mathcal{K}_{3,3}$) shown in Fig. 2.4 two possible sets of branch grouping are given as $\mathcal{P}(\mathcal{K}_{3,3}, S1) = \{BG(v_{1,1}), \ BG(v_{1,2}), \ BG(v_{1,3})\}$ and $\mathcal{P}(\mathcal{K}_{3,3}, S2) = \{BG(v_{2,1}), \ BG(v_{2,2}), \ BG(v_{2,3})\}$, where the involved branch groups are $BG(v_{1,j}) = \{E_{j,1}, \ E_{j,2}, \ E_{j,3}\}$ and $BG(v_{2,k}) = \{E_{1,k}, \ E_{2,k}, \ E_{3,k}\}$ $(j, k = 1, 2, 3)$.

Specifically speaking, the concept of branch grouping can be utilized from the following two aspects: 1) Branch voltage transformation. The branch voltage sum from the same branch group contributes to control the external current at the shared terminal, which is applied later to decouple the MMCC model and deduce its dq model. 2) Branch energy balancing. The IBEB modeling utilizes the information of branch grouping that transforms the coupled individual branch energy regulation into a structural balancing strategy, which is balancing among the branch groups and inside the branch group. These two utilization will be explained in detail in the next chapter.

2.4 Summary

In this chapter, MMCC are explained from its fundamental cell (submodule), basic component (branch), branch-configured topology, topology-submodule-based structure to a complete MMCC system with external systems. Due to the fact that a large number of converter members in the MMCC family, two classic classification methods are reviewed. One classifies MMCCs by the conversion types, which cannot well distinguish the circuit configurations of MMCCs, especially for direct AC/AC MMCCs. Another method categorizes MMCC topologies according to their three-phase connections, but it is limited to the classification of 3-phase MMCCs. Therefore, an analytical method based on graph theory is proposed to abstract the topological characteristics of MMCCs and classifies them with respect to the graph types. Additionally, with the help of graph analysis, each terminal and each branch are symbolized under a unified rule, and each MMCC structure is identified by its graph type and submodule type. In the following chapter, the unified modeling framework for MMCC systems is presented based on this formal description of graph-based analysis.

3 Unified framework for modeling and analysis of MMCC systems

An MMCC system is a highly complex power electronics system including at least one MMCC structure and various external systems. During its operation continuous branch currents containing complex frequency components flow through the multiple converter branches. Furthermore, it is desirable to control its numerous submodule capacitor voltages in a dynamic stability. Therefore, this chapter develops an MMCC model that is capable of 1) describing the operation states of the whole MMCC system both in dynamics and in steady state, 2) achieving a compromise between accuracy and complexity and 3) facilitating the further model-based control design.

The general modeling of an MMCC system is handled in a hierarchical structure divided into three layers: external current dynamic modeling for MMCC functionality, internal branch energy modeling for MMCC branches and capacitor voltage modeling of the branch submodules. The presented unified framework in Part I focuses on the first two layers, and the third layer for submodule voltage balancing is realized by a valid sorting algorithm without requiring explicit submodule models. The layered analytical structure is shown in Fig. 3.1.

3.1 MMCC continuous equivalent circuit

Consider an MMCC with a fixed number of cascaded submodules in each branch, a straightforward modeling is to choose the submodule switching states as control inputs and to select the branch currents and submodule capacitor voltages as state variables, so that a functional relation between the control inputs and the state variables is established. However, as the number of branch submodules increases, it results in an increasingly complicated model with massive discrete control inputs, hybrid system property and complex couplings. It is also an impossible mission to design model-based controller that provides explicit solutions and satisfies the real-time requirement. To tackle this issue, branch equivalent circuit is applied to analyze an MMCC branch as a whole.

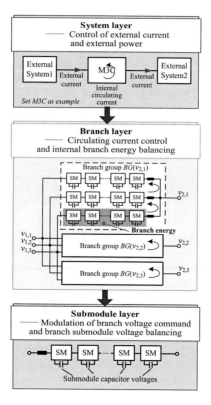

Figure 3.1: Layered analytical structure

3.1.1 Branch equivalent circuit

Investigate a general case for an MMCC branch $E_{j,k}$ with cascaded N_{SM} submodules as shown in Fig. 3.2(a). The N_{SM} branch submodules are assigned as SM l ($l = 1, 2, ..., N_{SM}$) with the corresponding switching state function g_l and the capacitor voltage $u_{c,l}$. The branch current and the branch voltage are labeled as $i_{j,k}$ and $u_{j,k}$, which are named after the symbolization $E_{j,k}$. All the cascaded submodules in the same branch are called branch submodules (BSMs). The symbol L is the branch inductance, and R is the equivalent branch resistance modeling the branch losses.

At each time instant t, the branch voltage $u_{j,k}(t)$ is defined as output voltage from

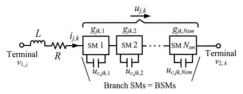

(a) Detailed switch circuit

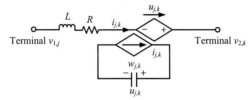

(b) CVCS-based branch equivalent circuit

Figure 3.2: Switched and continuous equivalent branch models

the branch submodules. Note that the voltage drop across the branch inductor is not counted in the branch voltage.

$$u_{j,k}(t) = \sum_{l=1}^{N_{SM}} g_{jk,l}(t) u_{c,jk,l}(t) \tag{3.1}$$

where the switching functions are defined as

$$g_{jk,l}(t) \in \begin{cases} \{0,\ 1\} & \text{for a HB-SM} \\ \{-1,\ 0,\ 1\} & \text{for a FB-SM} \end{cases} \tag{3.2}$$

As already roughly explained before, the above equation (3.1) may contain a large number of submodule switching states, which are dynamically described by (3.2). It therefore results in a large-scaled hybrid constrained system, which is impossible neither to develop an explicit controller nor to be efficiently solved online. In engineering practices such straightforward and complicated MMCC models are not advisable, especially for the case of MMCC with a high system dimension. The continuous equivalent analysis is required to develop a simple and effective model for MMCC, which begins with branch equivalent circuit.

To model the MMCC branch, the equivalent CVCS model is introduced to replace the cascaded submodules, which integrates the branch current dynamics and the branch energy dynamics. Compared with the simplest equivalence by replacing the BSMs with a single voltage source, CVCS-based equivalent method establishes a relation between the branch voltage and the limited energy storage in MMCC capacitors [42]. The CVCS equivalence can be applied for each submodule to obtain per-submodule model or for the whole branch to obtain per-branch model. In order to achieve a generalized branch modeling regardless of the quantity of branch submodule, the per-branch model is selected as shown in Fig. 3.2(b).

The functionality of the CVCS model can be explained as follows. As the branch submodules are switched on and off, branch voltage $u_{j,k}(t)$ changes accordingly, forming the controllable voltage source. After that, the controllable voltage source interacts with the terminal voltage to determine the branch current $i_{j,k}(t)$. The controllable current source is directly related to $i_{j,k}(t)$ and the joint interaction of $i_{j,k}(t)$ and $u_{j,k}(t)$ contributes to change the total energy stored in branch submodules. This total energy is termed as the branch energy, which is defined as:

$$w_{j,k}(t) = \sum_{l=1}^{N_{SM}} (\frac{1}{2}Cu_{c,jk,l}^2(t)) \tag{3.3}$$

where C is the submodule capacitance, $u_{c,jk,l}$ is the capacitor voltage of lth submodule and N_{SM} is the quantity of branch submodules. Although the branch submodules are switched on and off during operation, the branch energy $w_{j,k}$ is evaluated by the sum of the instantaneous submodule capacitor voltages regardless of their instantaneous switching states. The deduction of (3.3) can be found in the Appendix A

3.1.2 MMCC model integrating with current and energy aspects

With the help of CVCS-based branch model the MMCC system is analyzed by integrating current and energy dynamics, which develops the integrated MMCC model. As shown in Fig. 3.3, the general diagram of an MMCC system is decomposed into three links, which are the external link for System1, MMCC internal link and the external link for System2. For the MMCC external links, the equivalent electrical circuits are established by external voltage sources, external passive components and MMCC external control voltages. The MMCC external control voltage is produced by a linear combination of the branch voltages. From the external link, the external equivalent circuit is developed, which is applied to analyze the external current behaviors as well as the delivered external power. Between the both-side external links there exists an MMCC internal link. The internal equivalent circuit for the internal link is applied to analyze the behavior of MMCC internal currents that flow inside the MMCC and regulate energy distribution among the converter branches. Such internal currents exist widely in

MMCC topologies, which are commonly termed as circulating currents (CC). The three links are bonded by conforming to the energy conservation law.

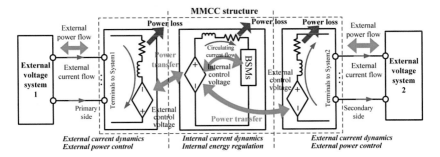

Figure 3.3: Schematic diagram of current dynamic and energy transfer in MMCC system

The above mentioned external currents, circulating currents, external power flow and internal branch energy are the four modeling objects in the proposed unified modeling framework.

3.1.3 Unified modeling framework

The unified modeling framework allows analysis and modeling of MMCC systems with its procedures given in Fig. 3.4. This framework aims to develop MMCC models in a hierarchical way regardless of its structure type and connected external systems. As indicated in Fig. 3.4, the entire MMCC modeling task is divided into current layer (the blue block) and energy layer (the green block). Each layer is then decoupled into external scope (the orange block) and internal scope (the red block). The final developed model is formulated in a standard state-space representation oriented towards the future multivariable control design. The following two sections 3.2 and 3.3 will illustrate the two modeling layers in detail, respectively.

3.2 Current dynamic modeling of MMCC system

A complete state-space description of MMCC current dynamics will firstly be developed in this section, which is composed of the external current dynamics and the CC dynamics.

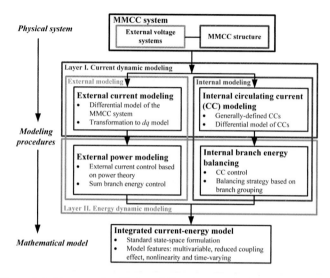

Figure 3.4: Diagram of unified modeling framework of MMCC systems

The basic configuration of the obtained current dynamic model is given in advance as

$$\frac{d}{dt}\begin{pmatrix} i_s \\ i_z \end{pmatrix} = \begin{pmatrix} A_t & 0 \\ 0 & A_z \end{pmatrix}\begin{pmatrix} i_s \\ i_z \end{pmatrix} + \begin{pmatrix} B_t & 0 \\ 0 & B_z \end{pmatrix}\begin{pmatrix} u_s \\ u_z \end{pmatrix} + \begin{pmatrix} E_t \\ 0 \end{pmatrix} V_s \qquad (3.4)$$

where the index matrices involved with external current dynamics are indicated by the subscript "t" and the index matrices related to circulating current dynamics are symbolized by "z". The state variables are composed of the external currents i_s and the CCs i_z, and accordingly the control inputs comprise the external control voltages u_s and the internal control voltages u_z. $\left(i_s \;\; i_z \right)^T$ contains the selected independent current variables for fully describing the current dynamics of an MMCC system. $\left(u_s \;\; u_z \right)^T$ are determined by performing an equivalent transformation to the original N_{br} branch voltage variables u_{br}. The external source voltages V_s from the external voltage systems influence the external current dynamics. This section begins with development of external current model.

3.2.1 Branch current modeling

Firstly, the differential equation for one MMCC branch shown in Fig. 3.2(b) is formulated as

$$\frac{d}{dt}i_{j,k} = -\frac{R}{L}i_{j,k} - \frac{1}{L}u_{j,k} + \frac{1}{L}(V_{1,j} - V_{2,k}) \tag{3.5}$$

where $V_{1,j}$ and $V_{2,k}$ are the external voltages at $v_{1,j}$ and $v_{2,k}$. Similar differential equations can be listed for all the N_{br} branches and collected into a matrix form as

$$\frac{d}{dt}\boldsymbol{i}_{br} = \boldsymbol{A}_{br}\boldsymbol{i}_{br} + \boldsymbol{B}_{br}\boldsymbol{u}_{br} + \boldsymbol{E}_{br}\boldsymbol{V}_t \tag{3.6}$$

where $\boldsymbol{i}_{br}$ are the branch currents, $\boldsymbol{u}_{br}$ are the branch voltages and $\boldsymbol{V}_t$ are the external voltages at the terminals. (3.6) is the basis for developing the external and circulating current models.

3.2.2 External current modeling

According to Kirchhoff junction rule, one external current is the sum of the branch currents at this terminal. Therefore, the independent external currents are calculated by summing the differential equations of branch currents as

$$\frac{d}{dt}\boldsymbol{i}_s = \boldsymbol{A}_s\boldsymbol{i}_s + \boldsymbol{B}_s\boldsymbol{u}_{br} + \boldsymbol{E}_s\boldsymbol{V}_t \tag{3.7}$$

The state-space equation (3.7) describes the current dynamics of the MMCC structure without the external voltage systems, featured as a linear time-invariant (LTI) system. (3.7) will be extended in Subsection 3.2.4 by replacing the external voltages with the external source voltages in order to obtain the model of an MMCC system. Additionally, the modeling of the connected systems will be presented in Subsection 3.2.4.

3.2.3 Internal current modeling

Many defined "circulating currents" (CCs) make only mathematical sense, but they fail to illustrate the physical meaning behind and cannot be extended for other topologies. For example, in [43, 40], the two CCs are defined for MMC (HB-$\mathcal{K}_{3,2}$) as the difference of the branch currents from the same phase, which actually contains additionally the current component from the connected DC system. Furthermore, a strictly-formulated determination of CCs in other MMCC topologies is not proposed yet. In this section CC is defined providing a clear physical interpretation and valid for all the MMCCs.

Definition of circulating current
When an MMCC structure is connected to balanced external voltage systems at its terminals, the CCs are purely determined by the voltage difference among the converter branches. Under this condition, CC can be defined as

Definition 3.1. *Circulating current is the current that gives no contribution to the external current.*

Based on Definition 3.1 that no CC flows to the external systems contributing external currents, the following constraint equations can be obtained for each terminal:

$$\begin{cases} \sum_{k=1}^{N} i_{z,jk} = 0 & \text{for Terminal } v_{1,j} \\ \sum_{j=1}^{M} i_{z,jk} = 0 & \text{for Terminal } v_{2,k} \end{cases} \tag{3.8}$$

where $i_{z,jk}$ is the CC component flowing through Branch $E_{j,k}$. For an MMCC topology with totally N_{tm} terminals, (3.8) contains N_{tm} constraint equations, whereby $N_{tm}-1$ ones are independent. Since there exist N_{br} unknown CC components for the N_{br} branches, $(N_{br} - N_{tm} + 1)$ independent CC components can be solved out. They are collected into a vector $\boldsymbol{i}_z$, which serves for describing MMCC internal current dynamics.

CC expression based on branch currents and differential form
On the basis of independent CC components in an MMCC, the subsequent problem is to obtain their explicit mathematical expressions based on measurable branch currents and also the differential dynamic equations. This problem can be handled as follows:

1. The branch current $i_{j,k}$ can generally be expressed as the sum of the external current component $i_{s,jk}$ in this branch and the internal CC component $i_{z,jk}$, which is

$$i_{j,k} = i_{s,jk} + i_{z,jk} \tag{3.9}$$

   The above equation links branch current, external current component and CC component together. If the different equations of the external current component are obtained, the corresponding dynamic equations for CC components are also determined.

2. Due to the homogeneity and symmetry of branch configuration, the external current distributes equally among the branches connecting at the terminal. Therefore, the external current component $i_{s,jk}$ is calculated by the external current $i_{j,k}$ as

$$i_{s,jk} = f_{\mathcal{X}}(\boldsymbol{i}_s) \tag{3.10}$$

   where $f_{\mathcal{X}}$ is a linear function determined by the branch arrangement of MMCC $\mathcal{X}$. Note that the senses of the equal distribution are different for MMCC $\mathcal{K}_{M,N}$

and $\mathcal{C}_{M+N}$. For $\mathcal{K}_{M,N}$, the external current component from the branches in the same branch group share the corresponding external current with the same phase angle but the equally-divided amplitude, because of the parallel-connected branch configuration of $\mathcal{K}_{M,N}$. On the contrast, for MMCC $\mathcal{C}_{M+N}$, the external current components from the branches in the same branch group share their external current with the same amplitude but shifted phase angles ($\pm 60°$).

3. As already illustrated before, each CC component $i_{z,jk}$ can be described by the selected independent $\boldsymbol{i}_z$ as

$$i_{z,jk} = g_{\mathcal{X}}(\boldsymbol{i}_z) \tag{3.11}$$

$g_{\mathcal{X}}$ is also a linear function with respect to the specific MMCC topology.

4. Combine (3.9), (3.10) and (3.11), the branch current can generally be expressed by external current components and independent CC component as

$$i_{j,k} = f_{\mathcal{X}}(\boldsymbol{i}_s) + g_{\mathcal{X}}(\boldsymbol{i}_z) \tag{3.12}$$

5. By listing the analogous equation (3.12) for all the N_{br} branches and canceling the external current terms $f_{\mathcal{X}}(\boldsymbol{i}_s)$, the CCs $\boldsymbol{i}_z$ can be fully described by the branch currents as

$$\boldsymbol{i}_z = h_{\mathcal{X}}(\boldsymbol{i}_{br}) \tag{3.13}$$

6. Refer to the equation set (3.6), the differential equations for the CCs can therefore be developed as

$$\frac{d}{dt}\boldsymbol{i}_z = \boldsymbol{A}_z \boldsymbol{i}_z + \boldsymbol{B}_z \boldsymbol{u}_{br} \tag{3.14}$$

Note the external voltage term $\boldsymbol{V}_t$ is canceled due to the condition that the external voltage systems are symmetric and balanced.

The developed CC model (3.14) is influenced only by the branch voltages $\boldsymbol{u}_{br}$. Regardless of the connected external systems, MMCC internal dynamic equation (3.14) remains the same for a determined MMCC. Appendix B gives the expressions of the CCs for different MMCC topologies.

3.2.4 External current modeling in dq frames

Consider the general case of an MMCC interconnecting two AC systems, each converter branch then connects two terminals with different fundamental frequencies, whereby the branch voltages contain the voltage components from the both frequencies. Consequently, an immediate application of the Park transformation to (3.7) encounters difficulty in dealing with the branch voltage dominated by multiple frequencies. This problem can be solved by following the steps:

1. Branch voltage transformation based on branch grouping. The branch voltage $\boldsymbol{u}_{br}$ can be transformed by an invertible transformation matrix $\boldsymbol{T}_{u,\chi}$ into three parts, external control voltages $\boldsymbol{u}_s$ based on (3.7), internal control voltages $\boldsymbol{u}_z$ based on (3.14) and the difference voltage u_{N0} between the two neutral points N and O of the external systems, which is

$$\boldsymbol{u}_{br} = \boldsymbol{T}_{u,\chi} \left(\boldsymbol{u}_s \quad \boldsymbol{u}_z \quad u_{N0}\right)^T \tag{3.15}$$

   $\boldsymbol{T}_{u,\chi}$ is a full-rank matrix containing only constant values without involving any electrical parameters. The submatrix in $\boldsymbol{T}_{u,\chi}$ can be formulated based on the results of branch grouping, whereby the sum of branch voltages from the same branch group determines the external current at this terminal. The inverse matrix $\boldsymbol{T}_{u,\chi}^{-1}$ presents a linear combination of the branch voltages. It should be mentioned that u_{N0} is required for an equivalent branch voltage transformation without loss of degree of freedom.

2. Modeling of external voltage systems. The external voltage $\boldsymbol{V}_t$ in (3.7) are expressed by the current dynamic models of the external voltage systems as

$$\boldsymbol{V}_t = \boldsymbol{f}_t(\boldsymbol{V}_s, \dot{\boldsymbol{i}}_s, \boldsymbol{i}_s, u_{N0}) \tag{3.16}$$

   where $\boldsymbol{V}_s$ is the modelled external source voltages. The function $\boldsymbol{f}_t()$ includes the external electrical parameters.

3. Final current dynamic model. By substituting the branch voltage transformation (3.15) and the external voltage expressions (3.16) into the current dynamic equations (3.7) and (3.14), a standard state-space representation of the MMCC system can be derived as shown in (3.4).

4. dq model. Since the state variables $\boldsymbol{i}_s$, the decoupled external control voltage $\boldsymbol{u}_s$ and also the disturbances $\boldsymbol{V}_s$ in (3.4) contain only single frequency components, a direct Park transformation can be applied to obtain the dq model of MMCC system.

The current dynamic model (3.4) in the abc frame is a linear time-invariant system. However, its dq model is linear time-varying (LTV) system, whereby its control index submatrix $\boldsymbol{B}_{z(dq)}$ contains time-varying terms related to the electrical angles of the connected external AC systems, denoted as φ_1 and φ_2.

3.2.5 Example of current model development for MMCC-$\mathcal{K}_{3,3}$

A case study for the branch voltage transformation and dq model is given by taking MMCC-$\mathcal{K}_{3,3}$ as example. The circuit configuration is shown in Fig. 3.5. For a clear illustration, the branch groups $BG(v_{1,1})$ and $BG(v_{1,2})$ are plotted in green color, and $BG(v_{2,1})$ and $BG(v_{2,2})$ are in yellow color.

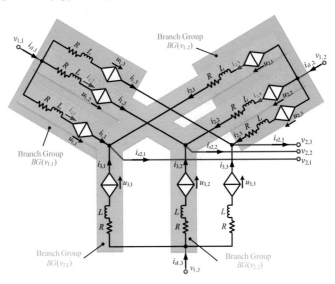

Figure 3.5: Illustration of $\mathcal{K}_{3,1}$-based MMCC $\mathcal{K}_{3,3}$

External current analysis
The differential equations for the branch currents are obtained based on Kirchhoff equations. Then the differential equations for the external currents are formulated by the branch current equations and the KCL equations at the terminals, which are

$$
\begin{cases}
\frac{d}{dt}i_{s1,1} = -\frac{R}{L}i_{s1,1} + \frac{1}{L}(u_{1,1} + u_{1,2} + u_{1,3}) + \frac{3}{L}V_{1,1} \\[2mm]
\frac{d}{dt}i_{s1,2} = -\frac{R}{L}i_{s1,2} + \frac{1}{L}(u_{2,1} + u_{2,2} + u_{2,3}) + \frac{3}{L}V_{1,2} \\[2mm]
\frac{d}{dt}i_{s1,3} = -\frac{R}{L}i_{s1,3} + \frac{1}{L}(u_{3,1} + u_{3,2} + u_{3,3}) + \frac{3}{L}V_{1,3} \\[2mm]
\frac{d}{dt}i_{s2,1} = -\frac{R}{L}i_{s2,1} + \frac{1}{L}(u_{1,1} + u_{2,1} + u_{3,1}) + \frac{3}{L}V_{2,1} \\[2mm]
\frac{d}{dt}i_{s2,2} = -\frac{R}{L}i_{s2,2} + \frac{1}{L}(u_{1,2} + u_{2,2} + u_{3,2}) + \frac{3}{L}V_{2,2}
\end{cases}
\tag{3.17}
$$

Write into a compact matrix form as

$$\frac{d}{dt}\boldsymbol{i}_s = \boldsymbol{A}_s\boldsymbol{i}_s + \boldsymbol{B}_s\boldsymbol{u}_{br} + \boldsymbol{E}_s\boldsymbol{V}_t \tag{3.18}$$

where $\boldsymbol{i}_s = \begin{pmatrix} i_{s1,1} & i_{s1,2} & ... & i_{s2,3} \end{pmatrix}^T$ are the M3C external currents, $\boldsymbol{u}_{br} = \begin{pmatrix} u_{1,1} & ... & u_{3,3} \end{pmatrix}^T$ are the branch voltages and $\boldsymbol{V}_t = \begin{pmatrix} V_{1,1} & V_{1,2} & ... & V_{2,3} \end{pmatrix}^T$ are its terminal voltages. The index matrices $\boldsymbol{A}_s$, $\boldsymbol{B}_s$ and $\boldsymbol{E}_s$ are constant matrices involving the branch parameters R and L.

Circulating current analysis

Based on the equations in (B.3) and the differential equations of the branch currents, the differential equations for the circulating currents are formulated as

$$\begin{cases} \frac{d}{dt}i_{z,1} &= -\frac{R}{L}i_{z,1} + \frac{1}{3L}(u_{1,1} - u_{1,2} - u_{1,3} - u_{2,1} - u_{3,1}) \\ \frac{d}{dt}i_{z,2} &= -\frac{R}{L}i_{z,2} + \frac{1}{3L}(-u_{1,1} + u_{1,2} - u_{1,3} - u_{2,2} - u_{3,2}) \\ \frac{d}{dt}i_{z,3} &= -\frac{R}{L}i_{z,3} + \frac{1}{3L}(-u_{1,1} + u_{2,1} - u_{2,2} - u_{2,3} - u_{3,1}) \\ \frac{d}{dt}i_{z,4} &= -\frac{R}{L}i_{z,4} + \frac{1}{3L}(-u_{1,2} - u_{2,1} + u_{2,2} - u_{2,3} - u_{3,2}) \end{cases} \tag{3.19}$$

They can be written in an alternative form as

$$\frac{d}{dt}\boldsymbol{i}_z = \boldsymbol{A}_z\boldsymbol{i}_z + \boldsymbol{B}_z\boldsymbol{u}_{br} \tag{3.20}$$

where $\boldsymbol{i}_z = \begin{pmatrix} i_{z,1} & i_{z,2} & i_{z,3} & i_{z,4} \end{pmatrix}^T$. $i_{z,1}$, $i_{z,2}$, $i_{z,3}$ and $i_{z,4}$ correspond to the CC components in Branch $E_{1,1}$, $E_{1,2}$, $E_{2,1}$ and $E_{2,2}$. $\boldsymbol{A}_z$ and $\boldsymbol{B}_z$ are constant matrices.

Current dynamic modeling in abc frame

The external modeling of the M3C system is considered as shown in Fig. 3.6. The external connections can generally be modeled as a back EMF type load through series inductance and resistance.

The external voltage $\boldsymbol{V}_t$ in (3.18) can be replaced by the connected external voltage sources $\boldsymbol{V}_s = \begin{pmatrix} e_{s1,1} & e_{s1,2} & ... & e_{s2,3} \end{pmatrix}^T$ as:

$$\boldsymbol{V}_t = \boldsymbol{V}_s + \boldsymbol{u}_N - \boldsymbol{R}_{AC}\boldsymbol{i}_s - \boldsymbol{L}_{AC}\frac{d}{dt}\boldsymbol{i}_s \tag{3.21}$$

where $\boldsymbol{u}_N = \begin{pmatrix} 0 & 0 & 0 & u_{N0} & u_{N0} \end{pmatrix}^T$, u_{N0} is the potential difference between the two neutral points. $\boldsymbol{R}_{AC}$ and $\boldsymbol{L}_{AC}$ represent the resistive and inductive parameters of M3C

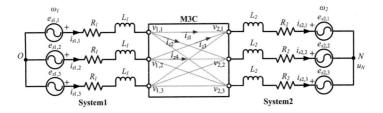

Figure 3.6: Generalized model of M3C with connected 3-phase systems

AC sides. Combine with (3.20), the model of the M3C system with external connections is given as

$$\frac{d}{dt}\begin{pmatrix} i_s \\ i_z \end{pmatrix} = \begin{pmatrix} A_{t(abc)} & 0 \\ 0 & A_z \end{pmatrix}\begin{pmatrix} i_s \\ i_z \end{pmatrix} + \begin{pmatrix} B_{t(abc)} \\ B_z \end{pmatrix} u_{br} + \begin{pmatrix} E_{t(abc)} \\ 0 \end{pmatrix} V_s \qquad (3.22)$$

Branch-grouping-based branch voltage transformation
It is concluded that the sum of the three branch voltages in the same branch group achieves control of the corresponding external currents, termed as u_s.

$$u_s = (u_{s1}\ u_{s2})^T = (u_{s1(1)}\ u_{s1(2)}\ u_{s2(1)}\ u_{s2(2)})^T \qquad (3.23)$$

where

$$\begin{cases} u_{s1(i)} & = \sum_{l=1}^{3} u_{i,l} \quad \text{for } BG(v_{1,i}) \text{ with } i = 1,\ 2 \\ u_{s2(j)} & = \sum_{l=1}^{3} u_{l,j} \quad \text{for } BG(v_{2,j}) \text{ with } j = 1,\ 2 \end{cases} \qquad (3.24)$$

The internal control voltage u_z for regulating CCs in (3.20) is given as

$$\begin{cases} u_{z1} & = u_{1,1} - u_{1,2} - u_{1,3} - u_{2,1} - u_{3,1} \\ u_{z2} & = -u_{1,1} + u_{1,2} - u_{1,3} - u_{2,2} - u_{3,2} \\ u_{z3} & = -u_{1,1} + u_{2,1} - u_{2,2} - u_{2,3} - u_{3,1} \\ u_{z4} & = -u_{1,2} - u_{2,1} + u_{2,2} - u_{2,3} - u_{3,2} \end{cases} \qquad (3.25)$$

so that $u_z = (u_{z,1}\ u_{z,2}\ u_{z,3}\ u_{z,4})^T$. By including the ninth equation that describes the potential difference u_{N0} that

$$u_{N0} = \frac{1}{9}\sum_{j=1}^{3}\sum_{k=1}^{3} u_{j,k} \qquad (3.26)$$

A linear transformation matrix $\boldsymbol{T}_{u,K33}$ can be determined for transforming the original control input $\boldsymbol{u}_{br}$ to the new ones as

$$
\begin{pmatrix} u_{s1(1)} \\ u_{s1(2)} \\ u_{s2(1)} \\ u_{s2(2)} \\ u_{z,1} \\ u_{z,2} \\ u_{z,3} \\ u_{z,4} \\ u_{N0} \end{pmatrix} = \begin{pmatrix} 1 & 1 & 1 & 0 & 0 & 0 & 0 & 0 & 0 \\ 0 & 0 & 0 & 1 & 1 & 1 & 0 & 0 & 0 \\ 1 & 0 & 0 & 1 & 0 & 0 & 1 & 0 & 0 \\ 0 & 1 & 0 & 0 & 1 & 0 & 0 & 1 & 0 \\ 1 & -1 & -1 & -1 & 0 & 0 & -1 & 0 & 0 \\ -1 & 1 & -1 & 0 & -1 & 0 & 0 & -1 & 0 \\ -1 & 0 & 0 & 1 & -1 & -1 & -1 & 0 & 0 \\ 0 & -1 & 0 & -1 & 1 & -1 & 0 & -1 & 0 \\ \frac{1}{9} & \frac{1}{9} & \frac{1}{9} & \frac{1}{9} & \frac{1}{9} & \frac{1}{9} & \frac{1}{9} & \frac{1}{9} & \frac{1}{9} \end{pmatrix} \begin{pmatrix} u_{1,1} \\ u_{1,2} \\ u_{1,3} \\ u_{2,1} \\ u_{2,2} \\ u_{2,3} \\ u_{3,1} \\ u_{3,2} \\ u_{3,3} \end{pmatrix} = \boldsymbol{T}_{u,K33}\boldsymbol{u}_{br} \qquad (3.27)
$$

(3.27) can be written in short as

$$
\boldsymbol{u}_{sz(9\times1)} = \begin{pmatrix} \boldsymbol{u}_{s(4\times1)} & \boldsymbol{u}_{z(4\times1)} & u_{N0} \end{pmatrix}^{T}_{(9\times1)} = \boldsymbol{T}_{u,K33(9\times9)}\boldsymbol{u}_{br(9\times1)} \qquad (3.28)
$$

where $\boldsymbol{T}_{u,K33}$ is a 9×9 square and invertible matrix that guarantees an equivalent transformation for the branch voltages.

Decoupled current dynamic modeling of M3C

As a summary, the current dynamic model (3.22) can be rewritten by substituting (3.28) as

$$
\frac{d}{dt}\begin{pmatrix} \boldsymbol{i}_s \\ \boldsymbol{i}_z \end{pmatrix} = \begin{pmatrix} \boldsymbol{A}_{t(abc)} & \boldsymbol{0} \\ \boldsymbol{0} & \boldsymbol{A}_z \end{pmatrix}\begin{pmatrix} \boldsymbol{i}_s \\ \boldsymbol{i}_z \end{pmatrix} + \boldsymbol{B}_{t(abc)}\boldsymbol{T}^{-1}_{u,M3C}\boldsymbol{u}_{sz} + \boldsymbol{E}_{t(abc)}\boldsymbol{V}_s \qquad (3.29)
$$

(3.29) is rewritten in a compact form as

$$
\frac{d}{dt}\boldsymbol{i}_{sz(abc)} = \boldsymbol{A}_{tz(abc)}\boldsymbol{i}_{sz(abc)} + \boldsymbol{B}_{tz(abc)}\boldsymbol{u}_{sz(abc)} + \boldsymbol{E}_{t(abc)}\boldsymbol{V}_{s(abc)} \qquad (3.30)
$$

The subscript (abc) in index matrices indicates the reference abc coordinate.

$\alpha\beta$ models of M3C systems

By introducing Clark transformation to state variables $\boldsymbol{x}_{sz(abc)}$, control inputs $\boldsymbol{u}_{sz(abc)}$ and disturbances $\boldsymbol{V}_{s(abc)}$ in(3.30), respectively, the $\alpha\beta$ components are written as (the zero component is assumed to be zero.)

$$
\begin{pmatrix} \boldsymbol{i}_{s(abc)} \\ \boldsymbol{i}_z \end{pmatrix} = \begin{pmatrix} \boldsymbol{T}_{2s/3s,is} & \boldsymbol{0} \\ \boldsymbol{0} & \boldsymbol{I} \end{pmatrix}\begin{pmatrix} \boldsymbol{i}_{s(\alpha\beta)} \\ \boldsymbol{i}_z \end{pmatrix} \quad \text{or} \quad \boldsymbol{i}_{sz(abc)} = \boldsymbol{T}_{2s/3s,i}\boldsymbol{i}_{sz(\alpha\beta)} \qquad (3.31)
$$

$$
\begin{pmatrix} \boldsymbol{u}_s \\ \boldsymbol{u}_z \\ u_{N0} \end{pmatrix} = \begin{pmatrix} \boldsymbol{T}_{2s/3s,us} & \boldsymbol{0} & \boldsymbol{0} \\ \boldsymbol{0} & \boldsymbol{I} & \boldsymbol{0} \\ \boldsymbol{0} & \boldsymbol{0} & 1 \end{pmatrix}\begin{pmatrix} \boldsymbol{u}_{s(\alpha\beta)} \\ \boldsymbol{u}_z \\ u_{N0} \end{pmatrix} \quad \text{or} \quad \boldsymbol{u}_{sz(abc)} = \boldsymbol{T}_{2s/3s,u}\boldsymbol{u}_{sz(\alpha\beta)} \qquad (3.32)
$$

$$V_{s(abc)} = T_{2s/3s,v} V_{s(\alpha\beta)} \tag{3.33}$$

where $T_{2s/3s,*}$ is the Clark transformation matrix that transfers from the abc frame to the $\alpha\beta$ frame for the corresponding variables $*$. By substituting the above relations to (3.30), it gives

$$T_{2s/3s,i}\frac{di_{sz(\alpha\beta)}}{dt} = A_{tz(abc)}T_{2s/3s,i}i_{sz(\alpha\beta)} + B_{tz(abc)}T_{u,\chi}^{-1}T_{2s/3s,u}u_{sz(\alpha\beta)} + E_{t(abc)}T_{2s/3s,v}V_{s(\alpha\beta)}$$

After a multiplication of $T_{2s/3s,i}^{-1}$ to the both sides, it gives

$$\frac{d}{dt}i_{sz(\alpha\beta)} = \underbrace{T_{2s/3s,i}^{-1}A_{tz(abc)}T_{2s/3s,i}}_{A_{tz(\alpha\beta)}}i_{sz(\alpha\beta)} + \underbrace{T_{2s/3s,i}^{-1}B_{tz(abc)}T_{u,\chi}^{-1}T_{2s/3s,u}}_{B_{tz(\alpha\beta)}}u_{sz(\alpha\beta)}$$

$$+ \underbrace{T_{2s/3s,i}^{-1}E_{t(abc)}T_{2s/3s,v}}_{E_{t(\alpha\beta)}}V_{s(\alpha\beta)}$$

Therefore, the $\alpha\beta$ model of an MMCC system is obtained as

$$\frac{d}{dt}i_{sz(\alpha\beta)} = A_{tz(\alpha\beta)}i_{sz(\alpha\beta)} + B_{tz(\alpha\beta)}u_{sz(\alpha\beta)} + E_{t(\alpha\beta)}V_{s(\alpha\beta)} \tag{3.34}$$

dq models of M3C systems
The Park transformation relations from variables in the $\alpha\beta$ frame to the dq frame are listed as

$$\begin{pmatrix} i_{s(\alpha\beta)} \\ i_z \end{pmatrix} = \begin{pmatrix} T_{2r/2s,is} & 0 \\ 0 & I \end{pmatrix} \begin{pmatrix} i_{s(dq)} \\ i_z \end{pmatrix} \quad \text{or} \quad i_{sz(\alpha\beta)} = T_{2r/2s,i}i_{s(dq)} \tag{3.35}$$

$$\begin{pmatrix} u_{s(\alpha\beta)} \\ u_z \\ u_{N0} \end{pmatrix} = \begin{pmatrix} T_{2r/2s,us} & 0 & 0 \\ 0 & I & 0 \\ 0 & 0 & 1 \end{pmatrix} \begin{pmatrix} u_{s(dq)} \\ u_z \\ u_{N0} \end{pmatrix} \quad \text{or} \quad u_{sz(\alpha\beta)} = T_{2r/2s,u}u_{sz(dq)} \tag{3.36}$$

$$V_{s(\alpha\beta)} = T_{2r/2s,v}V_{s(dq)} \tag{3.37}$$

where $T_{2r/2s,*}$ is the Park transformation matrix that obtains the dq components from the $\alpha\beta$ ones for the variables $*$. $T_{2r/2s,*}$ contains a time-varying angular variable φ that is the angle difference between the axis α and the axis d. Substitute the above relations to (3.34) as

$$
\begin{aligned}
\frac{dT_{2r/2s,i}i_{sz(dq)}}{dt} &= \frac{dT_{2r/2s,i}}{dt}i_{sz(dq)} + T_{2r/2s,i}\frac{di_{sz(dq)}}{dt} \\
&= A_{tz(\alpha\beta)}T_{2r/2s,i}i_{sz(dq)} + B_{tz(\alpha\beta)}T_{2r/2s,u}u_{sz(dq)} + E_{t(\alpha\beta)}T_{2r/2s,v}V_{s(dq)}
\end{aligned}
\tag{3.38}
$$

The above equation is further deduced as

$$\frac{d}{dt}\boldsymbol{i}_{sz(dq)} = \underbrace{(\boldsymbol{T}_{2r/2s,i}^{-1}\boldsymbol{A}_{tz(\alpha\beta)}\boldsymbol{T}_{2r/2s,i} - \boldsymbol{T}_{2r/2s,i}^{-1}\dot{\boldsymbol{T}}_{2r/2s,i})}_{\boldsymbol{A}_{tz(dq)}}\boldsymbol{i}_{sz(dq)} + \underbrace{\boldsymbol{T}_{2r/2s,i}^{-1}\boldsymbol{B}_{tz(\alpha\beta)}\boldsymbol{T}_{2r/2s,u}}_{\boldsymbol{B}_{tz(dq)}}\boldsymbol{u}_{sz(dq)}$$

$$+ \underbrace{\boldsymbol{T}_{2r/2s,i}^{-1}\boldsymbol{E}_{t(\alpha\beta)}\boldsymbol{T}_{2r/2s,v}}_{\boldsymbol{E}_{t(dq)}}\boldsymbol{V}_{s(dq)}$$

$$(3.39)$$

Note that $\boldsymbol{T}_{2r/2s,*}$ is a time-varying matrix dealing with φ_1 and φ_2. The solution of $\boldsymbol{T}_{2r/2s,i}^{-1}\dot{\boldsymbol{T}}_{2r/2s,i}$ is actually a rather simple constant matrix with ω_1 and ω_2.

As a conclusion, the dq model of the MMCC system can finally be given as

$$\frac{d}{dt}\boldsymbol{i}_{sz(dq)} = \boldsymbol{A}_{tz(dq)}\boldsymbol{i}_{sz(dq)} + \boldsymbol{B}_{tz(dq)}\boldsymbol{u}_{sz(dq)} + \boldsymbol{E}_{t(dq)}\boldsymbol{V}_{s(dq)} \qquad (3.40)$$

3.3 MMCC energy dynamic modeling

The second layer of the generalized modeling methodology is the energy dynamic modeling. Following the same division for the current modeling, the energy modeling can also be carried out respectively for the external scope and for the internal scope. The external scope is to model the external power that determines the external power transmission and the sum branch energy of an MMCC structure. The internal scope aims for IBEB that realizes a balanced energy distribution among the converter branches with the help of CC variables.

3.3.1 External power modeling

From the power aspect, the functionality of an MMCC structure is to deliver active power from one external system to another or to compensate the required reactive power. The external power can be regulated by the external currents $\boldsymbol{i}_s$ according to the power theory, for example, the instantaneous power theory or FBD theory.

After obtaining the external transmitted power, the next step is to model the sum branch energy. The sum branch energy w_Σ of an MMCC $\mathcal{X}$ is calculated by summing up all the branch energies $w_{j,k}$, which is

$$w_\Sigma = \sum_{E_{j,k} \in \mathcal{X}} w_{j,k} \qquad (3.41)$$

The distributed submodules inside MMCC can be treated as energy buffers for implementing energy conversion between external systems. According to the Law of Conservation of Energy, after subtracting the energy losses in the internal equivalent resistance, the power difference between the both-side external systems leads to charge or discharge submodule voltages simultaneously, so that the sum branch energy changes by following the equation below

$$\frac{d}{dt}w_\Sigma = p_{s1} - p_{s2} - p_{loss} = \boldsymbol{A}_{w\Sigma}\boldsymbol{i}_s - p_{loss} \tag{3.42}$$

where p_{s1} and p_{s2} are the external powers of the MMCC from System1 and System2 and p_{loss} is the power loss of the MMCC structure. p_{s1} and p_{s2} are from the multiplication of external voltage expression (3.16) and external currents, leading to the time-varying and nonlinear term $\boldsymbol{A}_{w\Sigma}\boldsymbol{i}_s$. The value of power loss p_{loss} is treated as disturbance calculated based on the developed current model at the operating point. The differential model (3.42) that controls the sum branch energy w_Σ is a nonlinear and time-varying equation, because the index matrix $\boldsymbol{A}_{w\Sigma}$ contains $\boldsymbol{i}_s$-dependent terms according to (3.16) and also time-varying $\boldsymbol{V}_s$.

The later proposed IBEB strategy aims to achieve a balanced energy distribution among all the converter branches with the controlled w_Σ. The following subsection illustrates the influence of $\boldsymbol{i}_s$ and $\boldsymbol{i}_z$ to the branch energy $w_{j,k}$.

3.3.2 Branch power expression

Refer to the differential equation of branch energy $w_{j,k}(t)$, it relates the branch power $p_{j,k}(t)$ as

$$\dot{w}_{j,k}(t) = -p_{j,k}(t) = -u_{j,k}(t)i_{j,k}(t) \tag{3.43}$$

where instantaneous branch power $p_{j,k}$ is calculated by the branch voltage $u_{j,k}$ and the branch current $i_{j,k}$. The minus sign indicates that a positive DC component in $p_{j,k}$ would reduce the branch energy $w_{j,k}$. It can be concluded that:

- A DC component in the branch power results in charging and discharging all the BSM capacitors to certain DC bias;

- AC components lead to the periodical voltage fluctuation across submodule capacitors but let the DC bias unchanged.

Then it is necessary to analyze the branch power expression (3.43) in detail so that the power fluctuation caused by the different frequency components and the corresponding control degrees of freedom can be inspected. Here FB-$\mathcal{K}_{3,3}$ connecting two AC systems with the frequencies f_1 and f_2 is taken as example. For an arbitrary branch $E_{j,k}$ ($E_{j,k} \in$

$\mathcal{K}_{3,3}$), the branch voltage $u_{j,k}(k)$ is approximated as the voltage difference between the two terminals and the branch current $i_{j,k}$ is expressed based on (3.12). By substituting these relations into (3.43) it gives

$$
\begin{aligned}
p_{j,k} =&\,(e_{s2,k} + u_{N0} - e_{s1,j})(\frac{1}{3}i_{s1,j} + \frac{1}{3}i_{s2,k} + i_{z,jk}) \\
=&\, -\frac{1}{3}\underbrace{e_{s1,j}i_{s1,j}}_{p_{s1,j}} + \frac{1}{3}\underbrace{e_{s2,k}i_{s2,k}}_{p_{s2,k}} + \underbrace{(-\frac{1}{3}e_{s1,j}i_{s2,k} + \frac{1}{3}e_{s2,k}i_{s1,j})}_{p_{s12,jk}} \\
&\, + \underbrace{u_{N0}(\frac{1}{3}i_{s1,j} + \frac{1}{3}i_{s2,k})}_{p_{N,jk}} - \underbrace{e_{s1,j}i_{z,jk}}_{p_{z(s1),jk}} + \underbrace{e_{s2,k}i_{z,jk}}_{p_{z(s2),jk}} + \underbrace{u_{N0}i_{z,jk}}_{p_{adj,jk}}
\end{aligned} \tag{3.44}
$$

Seven different power components can be observed in (3.44).

1. Injected power $p_{s1,j}$ from System1, Outflow power $p_{s2,k}$ to System2 and coupling power $p_{s12,jk}$: Due to the precondition of controlling external currents in a balanced sense, these power components change all the branch energy identically. Therefore, they cannot be used to adjust branch energy balancing.

2. Terminal-to-neutral power $p_{N,jk}$: This power component also relates to the external current, which cannot be used for balancing branch energy inside MMCC.

3. Circulating power $p_{z(s1),jk}$ and $p_{z(s2),jk}$: Certain frequency components can be introduced to the circulating current, which generates a DC bias in the circulating power and then changes the energy distribution among the targeted branch groups.

4. Adjacent power $p_{adj,jk}$, firstly proposed in [34] for Hexverter, originates from the both controllable variables u_{N0} and $i_{z,jk}$. Adjacent power is an important complement for realizing branch energy balancing. It also plays an important roll in ensuring a steady operation for a 3-phase/3-phase MMCC that connects two AC systems with identical or similar frequencies.

As a conclusion, the three power components, circulating power $p_{z(s1),jk}$, $p_{z(s2),jk}$ and adjacent power $p_{adj,jk}$, will produce an internal energy flow inside MMCC branches by introducing circulating current with specific amplitudes and frequencies. Due to the limited degrees of freedom and coupling effect, one branch energy cannot be altered without influencing the others. Thus, the balancing strategy adjusts the branch energy group by group, which is described in the following subsection.

3.3.3 Internal branch energy balancing

The Internal branch energy balancing (IBEB) strategy achieves the regulation of the sum branch energy by means of the external currents and the branch energy balancing by utilizing the internal flowing CCs. The functionality of CC is firstly explained.

Functionality of circulating current
The IBEB method is implemented based on the internal flowing CCs. Refer to (3.44), the functionality of regulating CCs can generally be explained for two different operating conditions.

1. In steady state operation, CCs should be eliminated or suppressed, while either its DC component or AC ones might increase the branch energy differences or unnecessarily cause additional submodule voltage ripples.

2. In dynamic operation, CCs are controlled according to (3.44) in order to achieve a preferred energy distribution among branches.

It is physically impossible to change only one branch energy without affecting the rest branches, due to the limited degrees of freedom of CCs and the operating requirement of balanced external currents. The proposed IBEB strategy aims to regulate the branch energy in groups, which are determined by the branch grouping method. Therefore, for an MMCC-$\mathcal{X}$ with N_{br} branches, the final IBEB goal is to determine correspondingly N_{br} independent energy variables and to develop their explicit dynamic equations based on CCs.

IBEB strategy

In the IBEB strategy the original N_{br} branch energies $\boldsymbol{w}_{br}$ of MMCC-$\mathcal{X}$ are not controlled individually to achieve the energy reference. Instead, another N_{br} energy variables $\boldsymbol{w}_{IBEB}$ will be deduced here, which are obtained by a linear equivalent transformation $\boldsymbol{T}_{w,\mathcal{X}}$ as

$$\boldsymbol{w}_{IBEB} = \boldsymbol{T}_{w,\mathcal{X}} \boldsymbol{w}_{br} \tag{3.45}$$

where the transformation matrix $\boldsymbol{T}_{w,\mathcal{X}}$ is an invertible constant matrix associated with the MMCC graph type $\mathcal{X}$. Fig. 3.7 shows the general procedure of the IBEB strategy. It is implemented in a layered structure by solving the balancing problem with the following three subgoals:

1. Subgoal 1: Control of sum branch energy
   The sum branch energy w_Σ of MMCC-$\mathcal{X}$ is calculated as the sum of all the branch

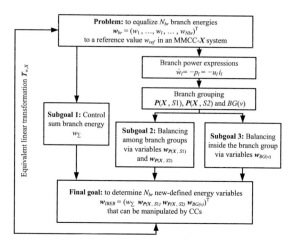

Figure 3.7: Procedure for IBEB design

energies

$$w_\Sigma = \sum_{j=1}^{M} \sum_{k=1}^{N} w_{j,k} \tag{3.46}$$

Then the first energy variable $w_{\Sigma,\Delta}$ is defined as the difference between w_Σ and $w_{\Sigma,ref} = N_{br} * w_{ref}$ that is

$$w_{\Sigma,\Delta} = w_\Sigma - w_{\Sigma,ref} \tag{3.47}$$

The differential form of $w_{\Sigma,\Delta}$ is expressed by the external currents and external voltages based on (3.42), which is written generally as

$$\dot{w}_{\Sigma,\Delta} = \boldsymbol{A}_{w\Sigma}\boldsymbol{i}_s - p_{loss} \tag{3.48}$$

where $\boldsymbol{A}_{w\Sigma}$ is a row vector containing MMCC external voltages as matrix elements.

2. Subgoal 2: Balancing among the branch groups
   The branches in an MMCC can be classified to a certain number of disjoint branch groups under Branch partitions $\mathcal{P}(\mathcal{X}, S1)$ and $\mathcal{P}(\mathcal{X}, S2)$. Compare to Subgoal 1 focusing on the overall energy level, Subgoal 2 focuses on the layer of branch groups. With the help of CCs, the energy distribution is adjusted group by group. Without loss of generality, the branch grouping of System1 $\mathcal{P}(\mathcal{X}, S1)$ is discussed here in detail.

The branch group energy $w_{s1(j)}$ of $BG(v_{1,j})$ in $\mathcal{P}(\mathcal{X}, S1)$ is calculated as

$$w_{s1(j)} = \sum_{(j,k) \in BG(v_{1,j})} w_{j,k} \qquad (3.49)$$

where $(j,k) \in BG(v_{1,j})$ indicates the branches $E_{j,k}$ in the branch group $BG(v_{1,j})$. For the partition $\mathcal{P}(\mathcal{X}, S1) = \{..., BG(v_{1,j}), ...\}$, the number of the branch groups depends on the number of primary-side terminals M (M varies from 2, 3, ...). The corresponding balancing strategy is also discussed with respect to M:

- If $M = 2$, two branch group energies $w_{s1(1)}$ and $w_{s1(2)}$ are determined. In order to evaluate their energy levels, one new energy variable is required by defining as the difference between them

$$w_{\mathcal{P}(s1),\Delta} = w_{s1(1)} - w_{s1(2)} \qquad (3.50)$$

- If $M = 3$, three branch group energies $w_{s1(1)}$, $w_{s1(2)}$ and $w_{s1(3)}$ are determined. Two new energy variables can be defined in a balanced sense to evaluate the differences among them, which are

$$\begin{cases} w_{\mathcal{P}(s1),\Delta 1} &= w_{s1(1)} - \frac{1}{2}(w_{s1(2)} + w_{s1(3)}) \\ w_{\mathcal{P}(s1),\Delta 2} &= w_{s1(2)} - \frac{1}{2}(w_{s1(1)} + w_{s1(3)}) \end{cases} \qquad (3.51)$$

- If M is an integer lager than 3, similarly $M - 1$ energy variables are defined to evaluate the differences among $\{w_{s1(1)}, w_{s1(2)}, ..., w_{s1(M)}\}$ that satisfies

$$w_{\mathcal{P}(s1),\Delta m} = w_{s1(m)} - \frac{1}{M-1} \sum_{l=1(l \neq m)}^{M-1} w_{s1(l)} \quad (m = 1, ..., M-1) \qquad (3.52)$$

Based on the branch power expression (3.43), the differential forms of the new energy variables in (3.50), (3.51) and (3.52) are calculated and simplified into a concise formulation only involving with the CC terms.

Note that $\mathcal{K}_{2,1}$, $\mathcal{K}_{3,1}$ and $\mathcal{C}_3$ are regarded as one branch group, although alternatively three branch groups with one single branch can also be observed. Their IBEB design for Subgoal 2 ought to be omitted.

3. Subgoal 3: Balancing inside one branch group
   At the bottom layer, the balancing issue is considered among the branches in the same branch group. As mentioned before, to balance N_{br} branch energies requires another independent N_{br} new variables. The number of the new energy variables

obtained in Subgoal 1 and 2 is fixed depending on the topology type. The number of missing independent variables will be determined in this layer. Similar to the balancing discussion for $\mathcal{P}(\mathcal{X}, S1)$, the formulation for $BG(v)$ can also be developed with respect to the branch number $n(BG(v))$.

- If $n(BG(v)) = 2$, one variable $w_{BG(v),\Delta}$ is defined as the difference between the two branch energies $w_{BG(v),1}$ and $w_{BG(v),2}$

$$w_{BG(v),\Delta} = w_{BG(v),1} - w_{BG(v),2} \qquad (3.53)$$

- If $n(BG(v)) = 3$, two variables $w_{BG(v),\Delta 1}$ and $w_{BG(v),\Delta 2}$ are defined as

$$\begin{cases} w_{BG(v),\Delta 1} &= w_{BG(v),1} - \frac{1}{2}\left(w_{BG(v),2} + w_{BG(v),3}\right) \\ w_{BG(v),\Delta 2} &= w_{BG(v),2} - \frac{1}{2}\left(w_{BG(v),1} + w_{BG(v),3}\right) \end{cases} \qquad (3.54)$$

By substituting the corresponding branch power expressions into (3.53) or (3.54) respectively, the differential forms of these new energy variables are developed with CCs and the potential difference, which is the adjacent power. The choice for choosing the branch groups is quite arbitrary as long as the necessary number for the missing energy variables is fulfilled.

IBEB synthesis

As shown in Fig. 3.7, $\boldsymbol{w}_{\mathcal{P}(\mathcal{X},S1)}$ and $\boldsymbol{w}_{\mathcal{P}(\mathcal{X},S2)}$ are the derived energy variables in Subgoal 2 and $\boldsymbol{w}_{BG(v)}$ are the ones in Subgoal 3, which are collected into the energy balancing term $\boldsymbol{w}_{bal} = \left(\boldsymbol{w}_{\mathcal{P}(\mathcal{X},S1)}\ \boldsymbol{w}_{\mathcal{P}(\mathcal{X},S2)}\ \boldsymbol{w}_{BG(v)}\right)^T$. Combined them with the sum energy w_{Σ}, exactly N_{br} new-defined energy variables $\boldsymbol{w}_{IBEB} = \left(w_{\Sigma}\ \boldsymbol{w}_{bal}\right)^T$ are complete to regulate the original N_{br} branch energies, which satisfies the form in (3.45). A differential form of $\boldsymbol{w}_{IBEB}$, which is also the dynamic IBEB model, is generally given as

$$\frac{d}{dt}\begin{pmatrix} w_{\Sigma} \\ \boldsymbol{w}_{bal} \end{pmatrix} = \begin{pmatrix} \boldsymbol{A}_{w\Sigma} & 0 \\ 0 & \boldsymbol{A}_{bal} \end{pmatrix}\begin{pmatrix} \boldsymbol{i}_s \\ \boldsymbol{i}_z \end{pmatrix} - \begin{pmatrix} p_{loss} \\ 0 \end{pmatrix} \qquad (3.55)$$

The multiplication of $\boldsymbol{A}_{bal}$ and $\boldsymbol{i}_z$ represents the circulating powers and the adjacent powers. Note that $\boldsymbol{A}_{bal}$ contains the external source voltages $\boldsymbol{V}_s$, which are time-varying terms from the external AC systems. Although the original coupling effects for regulation of the branch energies are eliminated in (3.55), the obtained dynamic model still contains time-varying terms due to the submatrices $\boldsymbol{A}_{bal}$ and nonlinearity due to $\boldsymbol{A}_{w\Sigma}$. An example of FB-$\mathcal{K}_{3,3}$ with a detailed description and the mathematical equations for this IBEB strategy is given subsequently.

3.3.4 Example of IBEB design for MMCC-$\mathcal{K}_{3,3}$

The M3C system depicted in Fig. 3.6 is considered here. The CCs flowing inside the M3C are used to balance the energy distribution among the nine branches. The equivalent partial circuit with controllable current sources is investigated here, and a general design procedure is followed according to Fig. 3.7 .

Subgoal 1: To control sum branch energy w_Σ
The sum branch energy w_Σ is defined as

$$w_\Sigma = \sum_{j=1}^{3} \sum_{k=1}^{3} w_{j,k} \tag{3.56}$$

with its reference value calculated as $w_{\Sigma,ref} = \frac{9}{2} N C_{SM} U^2_{SM,ref}$. Then the differential form for $w_{\Sigma,\Delta} = w_\Sigma - w_{\Sigma,ref}$ is given as

$$\dot{w}_{\Sigma,\Delta} = \sum_{k=1}^{3} e_{s2,k} i_{s2,k} - \sum_{j=1}^{3} e_{s1,j} i_{s1,j} - p_{loss} \tag{3.57}$$

By applying the decoupled power analysis in the two dq frames determined by external System1 and System2, the differential equation for sum branch energy is

$$\dot{w}_{\Sigma,\Delta} = \frac{3}{2}(e_{s2,d} i_{s2,d} + e_{s2,q} i_{s2,q}) - \frac{3}{2}(e_{s1,d} i_{s1,d} + e_{s1,q} i_{s1,q}) - p_{loss} \tag{3.58}$$

Subgoal 2: To achieve the balancing among the branch groups

- Balancing among $\mathcal{P}(\mathcal{K}_{3,3}, S1)$.
  For one branch group $BG(v_{1,j})$ in $\mathcal{P}(\mathcal{K}_{3,3}, S1)$, the sum energy of its branches $w_{s1(j)}$ is defined as
  $$w_{s1(j)} = w_{j,1} + w_{j,2} + w_{j,3} \tag{3.59}$$

Two additional energy variables $w_{\mathcal{P}(s1),\Delta1}$ and $w_{\mathcal{P}(s1),\Delta2}$ are defined in a balanced sense to evaluate the balancing states of the three branch-group energies as

$$\begin{cases} w_{\mathcal{P}(s1),\Delta1} = w_{s1(1)} - \frac{1}{2}(w_{s1(2)} + w_{s1(3)}) \\ w_{\mathcal{P}(s1),\Delta2} = w_{s1(2)} - \frac{1}{2}(w_{s1(1)} + w_{s1(3)}) \end{cases} \tag{3.60}$$

Their differential equations are derived based on the branch power expressions as

$$\begin{cases} \dot{w}_{\mathcal{P}(s1),\Delta1} = -\frac{3}{2}[e_{s2,1} i_{z1} + e_{s2,2} i_{z2} + e_{s2,3}(-i_{z1} - i_{z2})] \\ \dot{w}_{\mathcal{P}(s1),\Delta2} = -\frac{3}{2}[e_{s2,1} i_{z3} + e_{s2,2} i_{z4} + e_{s2,3}(-i_{z3} - i_{z4})] \end{cases} \tag{3.61}$$

A physical understanding of (3.61) is that f_2-frequency components in CCs are capable of adjusting the energy distribution among the three branch groups in $\mathcal{P}(\mathcal{K}_{3,3}, S1)$. Thus (3.61) can further be deduced by the f_2-frequency components $i_{z(s2)}$ as

$$
\begin{cases}
\dot{w}_{\mathcal{P}(s1),\Delta 1} &= -\frac{3}{2}[e_{s2,1}i_{z1(s2)} + e_{s2,2}i_{z2(s2)} + e_{s2,3}(-i_{z1(s2)} - i_{z2(s2)})] \\
\dot{w}_{\mathcal{P}(s1),\Delta 2} &= -\frac{3}{2}[e_{s2,1}i_{z3(s2)} + e_{s2,2}i_{z4(s2)} + e_{s2,3}(-i_{z3(s2)} - i_{z4(s2)})]
\end{cases}
\tag{3.62}
$$

Fig. 3.8 shows the principle of energy balancing under Partition $\mathcal{P}(\mathcal{K}_{3,3}, S1)$. The branches marked with the same color belong to the same branch group, and the circulating power has the identical influence to all of them. The orientation of $i_{z(s2)}$ determines the power flows $p_{z(s2),i}$ $i = 1$, 2, 3 (depicted as red, blue and green bars in Fig. 3.8) among the three branch groups, but their sum equals zero. The circulating currents can be oriented by the three-phase voltages of System2 and transformed into the dq frame for a decoupled control effect. The corresponding equations are

$$
\begin{cases}
\dot{w}_{\mathcal{P}(s1),\Delta 1} &= -\frac{9}{4}[e_{s2,d}i_{z12(s2),d} + e_{s2,q}i_{z12(s2),q}] \\
\dot{w}_{\mathcal{P}(s1),\Delta 2} &= -\frac{9}{4}[e_{s2,d}i_{z34(s2),d} + e_{s2,q}i_{z34(s2),q}]
\end{cases}
\tag{3.63}
$$

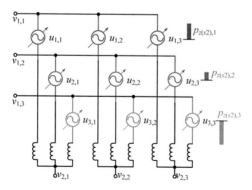

Figure 3.8: Illustration of branch energy balancing under Partition $\mathcal{P}(\mathcal{K}_{3,3}, S1)$

- Balancing among $\mathcal{P}(\mathcal{K}_{3,3}, S2)$.
  Two additional energy variables $w_{\mathcal{P}(s2),\Delta 1}$ and $w_{\mathcal{P}(s2),\Delta 2}$ are defined for the balancing purpose of the three branch groups in $\mathcal{P}(\mathcal{K}_{3,3}, S2)$ and their differential

equations can accordingly be obtained.

$$\begin{cases} \dot{w}_{\mathcal{P}(s2),\Delta 1} & = & -\frac{3}{2}[e_{s1,1}i_{z1} + e_{s1,2}i_{z3} + e_{s1,3}(-i_{z1} - i_{z3})] \\ \dot{w}_{\mathcal{P}(s2),\Delta 2} & = & -\frac{3}{2}[e_{s1,1}i_{z2} + e_{s1,2}i_{z4} + e_{s1,3}(-i_{z2} - i_{z4})] \end{cases} \tag{3.64}$$

Similarly, f_1-frequency components in $\boldsymbol{i}_z$ are used to manipulate the energy balancing among the three branch groups $BG(v_{2,1})$, $BG(v_{2,2})$ and $BG(v_{2,3})$ under partition $\mathcal{P}(\mathcal{K}_{3,3}, S2)$. It gives

$$\begin{cases} \dot{w}_{\mathcal{P}(s2),\Delta 1} & = & -\frac{3}{2}[e_{s1,1}i_{z1(s1)} + e_{s1,2}i_{z3(s1)} + e_{s1,3}(-i_{z1(s1)} - i_{z3(s1)})] \\ \dot{w}_{\mathcal{P}(s2),\Delta 2} & = & -\frac{3}{2}[e_{s1,1}i_{z2(s1)} + e_{s1,2}i_{z4(s1)} + e_{s1,3}(-i_{z2(s1)} - i_{z4(s1)})] \end{cases} \tag{3.65}$$

Fig. 3.9 shows the balancing mechanism similar as Fig. 3.8. (3.65) can also be expressed in the dq frame determined by System1 as

$$\begin{cases} \dot{w}_{\mathcal{P}(s2),\Delta 1} & = & -\frac{9}{4}[e_{s1,d}i_{z13(s1),d} + e_{s1,q}i_{z13(s1),q}] \\ \dot{w}_{\mathcal{P}(s2),\Delta 2} & = & -\frac{9}{4}[e_{s1,d}i_{z24(s1),d} + e_{s1,q}i_{z24(s1),q}] \end{cases} \tag{3.66}$$

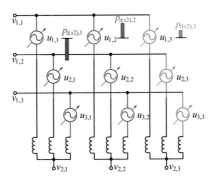

Figure 3.9: Illustration of branch energy balancing under Partition $\mathcal{P}(\mathcal{K}_{3,3}, S2)$

Subgoal 3: To achieve the balancing among the branches inside branch group
In this part, any two branch groups can be chosen to implement this subgoal.

- Balancing inside $BG(v_{1,1})$
  Two new energy variables are defined here to evaluate the balancing situation for

the three branch energies $w_{1,1}$, $w_{1,2}$ and $w_{1,3}$ as

$$\begin{cases} w_{BG(v1,1),\Delta1} & = & w_{1,1} - \frac{1}{2}(w_{1,2} + w_{1,3}) \\ w_{BG(v1,1),\Delta2} & = & w_{1,2} - \frac{1}{2}(w_{1,1} + w_{1,3}) \end{cases} \tag{3.67}$$

Their dynamic equations are expressed in a differential form as

$$\begin{cases} \dot{w}_{BG(v1,1),\Delta1} & = & -\frac{3}{2}u_{N0}i_{z1} \\ \dot{w}_{BG(v1,1),\Delta2} & = & -\frac{3}{2}u_{N0}i_{z2} \end{cases} \tag{3.68}$$

In the control practice the CCs with DC components contribute to adjust the balancing inside the branch group when the potential difference u_{N0} is regulated to a fixed DC offset. For this balancing task, DC components in i_{z1} and i_{z2} are demanded.

- Balancing inside $BG(v_{1,2})$

  The dynamic equations for the new defined energy variables are directly given here.

$$\begin{cases} \dot{w}_{BG(v1,2),\Delta1} & = & -\frac{3}{2}u_{N0}i_{z3} \\ \dot{w}_{BG(v1,2),\Delta2} & = & -\frac{3}{2}u_{N0}i_{z4} \end{cases} \tag{3.69}$$

It can be seen that DC components in i_{z3} and i_{z4} are required to adjust the balancing inside $BG(v_{1,2})$.

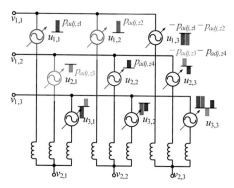

Figure 3.10: Illustration of branch energy balancing inside $BG(v_{1,1})$ and $BG(v_{1,2})$

Fig. 3.10 presents the balancing principle with DC components in CCs inside branch groups. Since there exist only four control degrees of freedom for the circulating power (depicted as red, green, blue and purple bars), a complete energy balancing among nine branches cannot be achieved. This subgoal is an important supplement to complete the IBEB strategy.

IBEB synthesis for M3C

The equivalent linear transformation matrix $\boldsymbol{T}_{w,M3C}$ from the nine branch energies $\boldsymbol{w}_{br}$ to another nine new energy variables $(w_\Sigma \;\; \boldsymbol{w}_{bal}^T)^T$ are formulated as $\left(w_\Sigma \;\; \boldsymbol{w}_{bal}^T\right)^T = \boldsymbol{T}_{w,M3C}\boldsymbol{w}_{br}$. The original problem of keeping the nine branch energies $\boldsymbol{w}_{br}$ to the reference value is now equivalently transformed to keep w_Σ to the reference and meanwhile regulate $\boldsymbol{w}_{bal}$ to zero with the help of the four CCs.

3.4 Integrated current-energy MMCC model

In this section, the current dynamic model (3.30) and the IBEB model (3.55) are combined to construct an integrated current-energy model for an MMCC system. The integrated model considers external current modeling, circulating current regulation, sum branch energy regulation and internal energy balancing, which is given as

$$\frac{d}{dt}\begin{pmatrix} \boldsymbol{i}_s \\ \boldsymbol{i}_z \\ w_\Sigma \\ \boldsymbol{w}_{bal} \end{pmatrix} = \begin{pmatrix} \boldsymbol{A}_t & \boldsymbol{0} & 0 & \boldsymbol{0} \\ \boldsymbol{0} & \boldsymbol{A}_z & 0 & \boldsymbol{0} \\ \boldsymbol{A}_{w\Sigma} & \boldsymbol{0} & 0 & \boldsymbol{0} \\ \boldsymbol{0} & \boldsymbol{A}_{bal} & 0 & \boldsymbol{0} \end{pmatrix}\begin{pmatrix} \boldsymbol{i}_s \\ \boldsymbol{i}_z \\ w_\Sigma \\ \boldsymbol{w}_{bal} \end{pmatrix} + \begin{pmatrix} \boldsymbol{B}_t & \boldsymbol{0} \\ \boldsymbol{0} & \boldsymbol{B}_z \\ \boldsymbol{0} & \boldsymbol{0} \\ \boldsymbol{0} & \boldsymbol{0} \end{pmatrix}\begin{pmatrix} \boldsymbol{u}_s \\ \boldsymbol{u}_z \end{pmatrix} + \begin{pmatrix} \boldsymbol{E}_t\boldsymbol{V}_s \\ \boldsymbol{0} \\ -p_{loss} \\ \boldsymbol{0} \end{pmatrix} \quad (3.70)$$

By inspecting the above integrated current-energy model, the system equations represent the following features from the control aspect:

Multi input - multi output (MIMO)

The MMCC system to be controlled always has more than one control input and more than one sensed output, which is concluded as a multivariable (multi-input, multi-output, or MIMO) system. The state-space design approach is selected throughout this thesis to describe the current dynamics and energy changes of MMCCs. Advantages of state-space design are especially obvious when dealing with MIMO systems, which are exactly the cases of MMCC systems.

Dynamical coupling effects

The index matrices in (3.70) are not diagonal matrices, so that the complicated coupling effects exist among the control inputs as well as among control of multiple system states. As there are no straightforward and one-to-one relationship between a control input and a state variable, it will increase difficulty of designing the traditional control structures, for example, the parallel or cascaded multi-PI controllers.

Time-variation

The index matrices for states and control inputs could contain time-varying terms introduced by the external voltage sources (φ_1 and φ_2). Fortunately, they are periodic variables with their respective periods T_1 and T_2. The time-varying characteristics should thus be handled by periodic modeling and periodic controller presented in Chapter 4.

Nonlinearity

A balanced internal energy distribution is always required for a safe long-term operation of an MMCC system in avoid of overcharging or over-discharging the branch submodules in some branches. However, by integrating the IBEB submodel into the complete MMCC current-energy model, it will also introduce nonlinearity through the power terms that originate from multiplications of branch currents and branch voltages.

As a conclusion, the proposed modeling method is developed based on the common characteristics of MMCCs and can guide for generating the mathematical models of all kinds of MMCC systems regardless their diverse topologies and different kinds of external systems. Based on the state-space model a control framework combined with multivariable systems and optimal principle will be employed to handle the aforementioned problems and to fulfill the control requirements.

3.5 Summary

CVCS-based branch equivalent model is selected to analyze the dynamic behaviors of MMCC structure, which is capable of taking the branch current and the averaging submodule voltage into account and still maintains the continuous feature of MMCC circuit variables. Based on that one MMCC system can be analyzed as a whole. A unified framework of MMCC modeling and analysis is built by following the procedures of external current modeling, internal circulating current modeling, external power modeling and internal branch energy balancing. The control of external current determines MMCC external functionality as well as the sum branch energy. By additionally producing certain circulating current components with specific amplitude and frequency, the energy distribution among MMCC branches can be adjusted group by group, so that a balanced status of branch energy is achieved. Since the proposed framework is generalized and valid for any MMCC structure, an automated algorithm based on the principle of Object-Oriented Modeling can be developed to generate the MMCC state-space representation.

4 Multivariable control methods for MMCC

In consideration of the major properties of the integrated MMCC model, e.g., MIMO, dynamical coupling, nonlinearity and time-variation, a control design based on conventional cascaded loops for separated control goals (external currents, internal currents, internal energies etc.) as shown in Fig. 4.1 would lead to a rather complex structure composed of dynamically interconnected subcontrollers, making the parameter tuning for those controllers a difficult job. Especially for fast transient requirements such as fault treatment the overall system behavior is difficult to be predicted and specifications can hardly be guaranteed. Furthermore, due to its specialty, if the MMCC structure or the external systems are changed, the layout of control structure has to be redesigned and the control parameters to be re-tuned. Such repetitive jobs actually can be avoided to the most extent by choosing an appropriate control strategy.

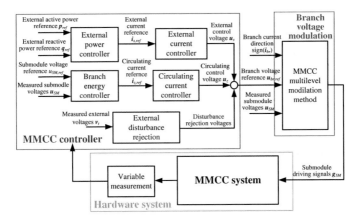

Figure 4.1: Diagram of cascaded MMCC control structure

4.1 Control design using state-space method

The state-space representation (3.70) can be expressed in a general way as

$$\dot{\boldsymbol{x}} = \boldsymbol{A}\boldsymbol{x} + \boldsymbol{B}\boldsymbol{u} + \boldsymbol{E}\boldsymbol{v} \qquad (4.1)$$

$$\boldsymbol{y} = \boldsymbol{C}\boldsymbol{x} + \boldsymbol{D}\boldsymbol{u} \qquad (4.2)$$

where system state $\boldsymbol{x}$ includes external currents, circulating currents, sum branch energy and IBEB variables, system input $\boldsymbol{u}$ contains the branch voltages, $\boldsymbol{v}$ is the modeled external voltage sources and $\boldsymbol{y}$ is the measured system variable. Without loss of generality, index matrices $\boldsymbol{A}$, $\boldsymbol{B}$, $\boldsymbol{E}$, $\boldsymbol{C}$ and $\boldsymbol{D}$ are set as constant, which indicate a LTI system. The further discussion will take the nonlinearity and time-variation into consideration. In the framework of digital control system, (4.1) is firstly discretized.

4.1.1 System discretization

The solution of (4.1) is given as

$$\boldsymbol{x}(t) = e^{\boldsymbol{A}t}\boldsymbol{x}(0) + \int_0^t e^{\boldsymbol{A}(t-\tau)}\boldsymbol{B}\boldsymbol{u}(\tau)\,d\tau \qquad (4.3)$$

By replace the time instants with kT to $kT+T$, a discrete recursive expression is derived as

$$\boldsymbol{x}(kT+T) = e^{\boldsymbol{A}T}\boldsymbol{x}(kT) + \int_{kT}^{kT+T} e^{\boldsymbol{A}(kT+T-\tau)}\boldsymbol{B}\boldsymbol{u}(\tau)\,d\tau \qquad (4.4)$$

The solution derived here is not dependent on the type of hold, because $\boldsymbol{u}(t)$ is specified in terms of its continuous time history $\boldsymbol{u}(\tau)$ over the sample interval. In the thesis zero-order holder (ZOH) without delay is applied, $\boldsymbol{u}(\tau)$ in (4.4) is supposed to be constant throughout the sample interval as

$$\boldsymbol{u}(\tau) = \boldsymbol{u}(kT),\ kT \le \tau < kT + T.$$

Thus, (4.4) can be transformed into

$$\boldsymbol{x}(kT+T) = e^{\boldsymbol{A}T}\boldsymbol{x}(kT) + \left(\int_0^T e^{\boldsymbol{A}\eta}\,d\eta\right)\boldsymbol{B}\boldsymbol{u}(kT) \qquad (4.5)$$

Let $\boldsymbol{\Phi} = e^{\boldsymbol{A}T}$ and $\boldsymbol{\Gamma} = \left(\int_0^T e^{\boldsymbol{A}\eta}\,d\eta\right)\boldsymbol{B}$. (4.1) and (4.2) are reduced to the difference equations as

$$\boldsymbol{x}(k+1) = \boldsymbol{\Phi}\boldsymbol{x}(k) + \boldsymbol{\Gamma}\boldsymbol{u}(k) + \boldsymbol{\Gamma}_1\boldsymbol{v}(k) \qquad (4.6)$$

$$\boldsymbol{y}(k) = \boldsymbol{C}\boldsymbol{x}(k) + \boldsymbol{D}\boldsymbol{u}(k) \tag{4.7}$$

Now a discrete representation for the continuous system is derived as (4.6) and (4.7). An important advantage of using state-space control design is that the control configuration can be separated into two parts. In the first step a full state feedback control law is deigned by assuming that all the system states are obtained at disposal. If not all the states are measurable, the second step aims to design an estimator to obtained these unmeasurable states. In an MMCC system the external currents, branch currents and submodule voltages can be acquired by the suitable sensors. The specific frequency components in circulating currents have to be estimated for control purpose.

4.1.2 Full state feedback control

As the way of rejecting the disturbances will be given later, the following discrete system model is analyzed by neglecting the disturbance term

$$\boldsymbol{x}(k+1) = \boldsymbol{\Phi}\boldsymbol{x}(k) + \boldsymbol{\Gamma}\boldsymbol{u}(k) \tag{4.8}$$

The control law is simply the product of system states $\boldsymbol{x}(k)$ and a control gain $\boldsymbol{K}$, that is

$$\boldsymbol{u}(k) = -\boldsymbol{K}\boldsymbol{x}(k) \tag{4.9}$$

Note that through the so-defined control law, the reference inputs $\boldsymbol{r}(k)$ are not included, which can be treated as $\boldsymbol{r}(k) = \boldsymbol{0}$. Therefore, the control law (4.9) is usually referred to as a regulator. The method to introduce the reference inputs is presented later. Substituting (4.9) in (4.8), it obtains

$$\boldsymbol{x}(k+1) = \boldsymbol{\Phi}\boldsymbol{x}(k) - \boldsymbol{\Gamma}\boldsymbol{K}\boldsymbol{x}(k) \tag{4.10}$$

The z-transform of the closed-loop system (4.10) is

$$(z\boldsymbol{I} - \boldsymbol{\Phi} + \boldsymbol{\Gamma}\boldsymbol{K})\boldsymbol{X}(z) = \boldsymbol{0} \tag{4.11}$$

The characteristic equation is

$$|z\boldsymbol{I} - \boldsymbol{\Phi} - \boldsymbol{\Gamma}\boldsymbol{K}| = 0 \tag{4.12}$$

By locating the n zeros of (4.12), that is, the n closed-loop eigenvalues to the desired locations, the stability and settling time of a closed-loop system is determined. Given desired pole locations inside the unit cycle $|z| = 1$ in the z plane as

$$z_i = \alpha_1,\ \alpha_2,\ \alpha_3,\ ...,\ \alpha_n$$

the desired control-characteristic equation is

$$g_c(z) = (z - \alpha_1)(z - \alpha_2)(z - \alpha_3) \ ... \ (z - \alpha_n) = 0 \tag{4.13}$$

(4.13) and (4.12) are the characteristic equations of the same controlled system, which are required identical term by term. Therefore, by equating the coefficients of the same power terms of z, there are n equations for a nth-order system and $m \times n$ unknowns. $(m-1) \times n$ of these knowns can be arbitrarily fixed. In general, there are many solutions to the pole assignment problem for MIMO systems. A canonical-form design method in [44] and many software tools can assist the task of pole placement. Although the mechanism of computing the control law is comparatively simple, the difficult and tricky task is to select an appropriate set of poles. LQR method shows its convenience of determining a stable control gain and advantage of handling high-order systems.

4.1.3 Estimator design

Since not all the system states need to be estimated, a reduced-order prediction estimator (ROPE) [44, 45] is employed to reconstruct the unmeasurable states. To illustrate ROPE a normal full-order prediction estimator is firstly introduced.

Prediction estimator
The predictive "open-loop" estimation of $\bar{x}(k)$ can be obtained based on the system model as

$$\bar{x}(k + 1) = \Phi \bar{x}(k) + \Gamma u(k) \tag{4.14}$$

The estimator error $\tilde{x}$ is defined as

$$\tilde{x}(k) = x(k) - \bar{x}(k) \tag{4.15}$$

Then the dynamic equation for the estimator error is expressed as

$$\tilde{x}(k + 1) = \Phi \tilde{x}(k) \tag{4.16}$$

The above "open-loop" estimator is quite simple in structure and easy to design. However, for a unstable and marginally stable plant, the error would never decreases from the initial value. Therefore, a compensation channel for the estimation error $\tilde{x}(k)$ multiplied by estimator gain L_p is constructed to constantly correct the estimated state $\bar{x}(k)$. The improved estimator equation with the correction channel is

$$\bar{x}(k + 1) = \Phi \bar{x}(k) + \Gamma u(k) + L_p[y(k) - C\bar{x}(k)] \tag{4.17}$$

After certain deduction, the behavior of estimation error $\tilde{x}(k)$ is described by the following difference equation as

$$\tilde{x}(k + 1) = [\Phi - L_p C]\tilde{x}(k) \tag{4.18}$$

The idea for designing the prediction estimator is to specify the desired estimator pole locations for the corresponding characteristic equation $|zI - \Phi + L_pC| = 0$ of (4.18) by adjusting the estimator gain L_p. The same method for designing the control law can be used here to determine L_p with the following estimation-characteristic equation

$$g_e(z) = (z - \beta_1)(z - \beta_2)(z - \beta_3) \ldots (z - \beta_n) = 0 \qquad (4.19)$$

where β_1, β_2, ..., β_n are the desired estimator pole locations. Consequently, the convergence rate for the estimated state vector $\bar{x}(k)$ converging toward the system state vector $x(k)$, correspondingly $\tilde{x}(k)$ converging to $\mathbf{0}$, can be determined.

4.1.4 Combined control law and estimator

It is proved that the substitution does not introduce unexpectable influence to the closed-loop control system [44]. The characteristic equation for such system is the product of the control- and estimator characteristic ones, which is

$$g_{cl} = g_c \cdot g_e = (z - \alpha_1)(z - \alpha_2) \ldots (z - \alpha_n)(z - \beta_1)(z - \beta_2) \ldots (z - \beta_n) \qquad (4.20)$$

where the poles of the closed-loop system consist of the combination of the control poles and the estimator poles. This explains that the control law and estimation can be design separately yet combined together, which is termed as the separation principal.

So far, the designed control law can simply regulate the system from any initial states to the origin according to the dynamics defined by the pole placement. The further discussion focuses on introduction of the reference input and rejection of the external disturbances.

4.1.5 Introduction of the reference input for MMCC

A control system usually guarantees a response to a non-zero reference in practical operation. It is necessary to introduce the reference input x_{ref} to the control system.

Suppose the system is in the steady state and the reference states x_{ref} are attained, (4.8) can be written as

$$x_{ss} = x_{ref} = \Phi x_{ref} + \Gamma u_{ref} \qquad (4.21)$$

A gain matrix N_u is imposed that $u_{ref} = N_u x_{ref}$, and (4.21) becomes

$$(\Phi - I)x_{ref} + \Gamma N_u x_{ref} = 0 \qquad (4.22)$$

$$(\Phi - I) + \Gamma N_u = 0 \qquad (4.23)$$

If the MMCC system does not have a zero at $z = 1$, $\boldsymbol{N}_u$ can be solved as

$$\begin{pmatrix} \boldsymbol{I} \\ \boldsymbol{N}_u \end{pmatrix} = \begin{pmatrix} \boldsymbol{\Phi} - \boldsymbol{I} & \boldsymbol{\Gamma} \\ \boldsymbol{I} & \boldsymbol{0} \end{pmatrix}^{-1} \begin{pmatrix} \boldsymbol{0} \\ \boldsymbol{I} \end{pmatrix} \tag{4.24}$$

Note that the influence from the state feedback channel should be subtracted from the steady-state control input $\boldsymbol{u}_{ref}$. The feedforward gain $\boldsymbol{N}_{ff}$ for reference input introduction is given as

$$\boldsymbol{N}_{ff} = \boldsymbol{N}_u + \boldsymbol{K} \tag{4.25}$$

Finally, the feedforward control law can be given as

$$\boldsymbol{u}_{ff} = \boldsymbol{N}_{ff} \boldsymbol{x}_{ref} \tag{4.26}$$

4.1.6 External disturbance rejection

A model including disturbance term is investigated here. Since the voltages from the external system can either be measured or estimated, disturbance rejection discussed here is classified as measurable disturbance rejection.

Consider the case that the disturbances enter on the input channel, where a matrix $\boldsymbol{M}$ exists such that $\boldsymbol{\Gamma}_1 = \boldsymbol{\Gamma M}$. When applying the control input that $\boldsymbol{u}(k) = -\boldsymbol{M v}(k)$, the disturbance can be cancelled as

$$\boldsymbol{x}(k+1) = \boldsymbol{\Phi x}(k) + \boldsymbol{\Gamma}(-\boldsymbol{M v}(k)) + \boldsymbol{\Gamma M v}(k) = \boldsymbol{\Phi x}(k) \tag{4.27}$$

where $\boldsymbol{x}(k)$ is independent of $\boldsymbol{v}(k)$. To verify if the conditions that $\boldsymbol{M}$ exists and $\boldsymbol{\Gamma}_1 = \boldsymbol{\Gamma M}$ satisfies, the following equation is considered.

$$\boldsymbol{\Gamma u}(k) = -\boldsymbol{\Gamma}_1 \boldsymbol{v}(k) \tag{4.28}$$

(4.28) should be solvable for $\boldsymbol{u}(k)$. The usual situation is that the dimension of $\boldsymbol{x}(k)$ is greater than the dimension of $\boldsymbol{u}(k)$. Then the least squares solution of (4.28) is given as

$$\boldsymbol{u}(k) = -(\boldsymbol{\Gamma}^T \boldsymbol{\Gamma})^{-1} \boldsymbol{\Gamma}^T \boldsymbol{\Gamma}_1 \boldsymbol{v}(k) \tag{4.29}$$

Perfect disturbance rejection can be achieved if $\boldsymbol{\Gamma}_1 = \boldsymbol{\Gamma}(\boldsymbol{\Gamma}^T \boldsymbol{\Gamma})^{-1} \boldsymbol{\Gamma}^T \boldsymbol{\Gamma}_1$ and if this is the case, it yields

$$\boldsymbol{M} = (\boldsymbol{\Gamma}^T \boldsymbol{\Gamma})^{-1} \boldsymbol{\Gamma}^T \boldsymbol{\Gamma}_1 \tag{4.30}$$

In the other cases, if $(\boldsymbol{I} - \boldsymbol{\Gamma}(\boldsymbol{\Gamma}^T \boldsymbol{\Gamma})^{-1} \boldsymbol{\Gamma}^T) \boldsymbol{\Gamma}_1$ is small enough compared to $\boldsymbol{\Gamma}_1$, an approximated rejection for external disturbance can also be achieved [46].

4.2 Linear Quadratic Regulator

In Section 4.1, a complete control design that uses the pole-placement is propose to deal with MIMO systems. The determination of the control gain K is actually quite flexible and of course not unique, which needs extra knowledge for understanding the MIMO systems and appropriate utilization of the extra freedom for them. So the selection of suitable pole locations can be a tricky task. In this section optimal control principle is introduced to assist the selection of control gains for MIMO systems.

The idea of multivariable optimal control can be described briefly as: a linear multivariable model for the investigated system is required in advance and a quadratic cost function J is defined based on the concerned control criteria. The cost function J achieves a compromise between the use of control effort and the response time of the state variables, which can inherently achieve a stable closed-loop system. A stable control law, e.g., K, can uniquely be determined after minimizing J.

4.2.1 Time-varying multivariable optimal control

Consider a discrete LTV model

$$x(k+1) = \Phi_k x(k) + \Gamma_k u(k) \tag{4.31}$$

with $x(0) = x_0$, a control law

$$u(k) = -K_k x(k) \tag{4.32}$$

can be so determined that the following cost function

$$J = \frac{1}{2} x^T(N) S x(N) + \frac{1}{2} \sum_{k=0}^{N-1} [x^T(k) Q_k x(k) + u^T(k) R_k u(k)] \tag{4.33}$$

is minimized. S is a symmetric and positive semi-definite weighting matrix for penalizing the end state $x(N)$, Q_k are symmetric and positive semi-definite weighting matrices, namely error weighted matrices, and R_k are symmetric positive definite weighting matrices, namely control weighted matrices, all of which are chosen by the designer to suit the application requirements. The optimal control $u^*(k)$ for $0 \leq k \leq N$ can be solved with dynamic programming method [47] and Lagrange multipliers [44]. Both the methods lead to the same solution that the optimal control as

$$u^*(k) = -K_k x(k) \tag{4.34}$$

at the time instant k, where the gain matrices K_k are given by

$$K_k = (R_k + \Gamma_k^T P_{k+1} \Gamma_k)^{-1} \Gamma_k^T P_{k+1} \Phi_k \tag{4.35}$$

and the symmetric positive semi-definite matrices $\boldsymbol{P}_k$ are determined recursively by

$$\boldsymbol{P}_k = \boldsymbol{\Phi}_k^T(\boldsymbol{P}_{k+1} - \boldsymbol{P}_{k+1}\boldsymbol{\Gamma}_k(\boldsymbol{R}_k + \boldsymbol{\Gamma}_k^T\boldsymbol{P}_{k+1}\boldsymbol{\Gamma}_k)^{-1}\boldsymbol{\Gamma}_k^T\boldsymbol{P}_{k+1})\boldsymbol{\Phi}_k + \boldsymbol{Q}_k \qquad (4.36)$$

where $\boldsymbol{P}_N = \boldsymbol{S}$. (4.36) is called the discrete-time Riccati equation and the control gains $\boldsymbol{K}_k$ are the linear time-varying feedback control gain matrices, which are called Kalman gains. Since the calculation of $\boldsymbol{P}_k$ and $\boldsymbol{K}_k$ does not require a given initial state $\boldsymbol{x}_0$, they can be computed offline and stored for the online acquisition.

4.2.2 LQR optimal control

The obtained control gains $\boldsymbol{K}_k$ in Subsection 4.2.1 is time-varying, but the early gains are exactly the same for a comparatively long time when the optimization length N is selected long enough. Therefore, a portion from the solution $\boldsymbol{K}_k$ can produce a constant gain $\boldsymbol{K}$ that is actually the solution for the infinite time case of time-varying optimal control. Such regulator that can regulate the state vector of a system from any initial state to the origin is called linear quadratic regulator, or LQR, because it regulates a LTI system with a quadratic form of the cost function. The cost function for the LQR case can be generally written as

$$J = \frac{1}{2}\sum_{k=0}^{\infty}[\boldsymbol{x}^T(k)\boldsymbol{Q}\boldsymbol{x}(k) + \boldsymbol{u}^T(k)\boldsymbol{R}\boldsymbol{u}(k)] \qquad (4.37)$$

where the horizon of the cost function is set to infinite and the weighting matrices $\boldsymbol{Q}$ and $\boldsymbol{R}$ are set to be constant.

Instead of solving the discrete-time Riccati equation (4.36) for time-varying control gains $\boldsymbol{K}_k$, the constant control gain $\boldsymbol{K}$ can be solved by the steady-state form of (4.36) that $\boldsymbol{P}_k$ becomes equal to $\boldsymbol{P}_{k+1}$. Suppose in the steady state, $\boldsymbol{P}_k = \boldsymbol{P}_{k+1} = \boldsymbol{P}$, then (4.36) can be written for a LTI system that $\boldsymbol{\Phi}_k = \boldsymbol{\Phi}$ and $\boldsymbol{\Gamma}_k = \boldsymbol{\Gamma}$ as

$$\boldsymbol{P} = \boldsymbol{\Phi}^T(\boldsymbol{P} - \boldsymbol{P}\boldsymbol{\Gamma}(\boldsymbol{R} + \boldsymbol{\Gamma}^T\boldsymbol{P}\boldsymbol{\Gamma})^{-1}\boldsymbol{\Gamma}^T\boldsymbol{P})\boldsymbol{\Phi} + \boldsymbol{Q} \qquad (4.38)$$

which is called the algebraic Riccati equation (ARE). The positive definite solution $\boldsymbol{P}$ for (4.38) can be calculated by numerical methods, which are supported by many software packages.

With the final calculated $\boldsymbol{P}$, the stationary control law for LQR is

$$\boldsymbol{u}(k) = -\boldsymbol{K}\boldsymbol{x}(k) = -(\boldsymbol{R} + \boldsymbol{\Gamma}^T\boldsymbol{P}\boldsymbol{\Gamma})^{-1}\boldsymbol{\Gamma}^T\boldsymbol{P}\boldsymbol{\Phi}\ \boldsymbol{x}(k) \qquad (4.39)$$

For LQR design, the selections of $\boldsymbol{Q}$ and $\boldsymbol{R}$ indicate the relative importance between convergence rate of system states and the demanded control effort, and the elements in the weighting matrices represent the specific concern to their internal elements. The tuning of $\boldsymbol{Q}$ and $\boldsymbol{R}$ requires a certain amount of trial and error in order to obtain a satisfactory control effect.

4.3 Nonlinear feedback control

In the integrated model (3.70) the partitioned matrix $\boldsymbol{A}_{bal}$ for IBEB contains the branch voltage variables, e.g., control input. A generalized formulation can be given as

$$\dot{\boldsymbol{x}}(t) = \boldsymbol{A}(\boldsymbol{u}(t))\boldsymbol{x}(t) + \boldsymbol{B}\boldsymbol{u}(t) \qquad (4.40)$$

which can be equivalently represented by the following linear structure with the state-dependent coefficients (SDC):

$$\dot{\boldsymbol{x}}(t) = \boldsymbol{A}\boldsymbol{x}(t) + \boldsymbol{B}(\boldsymbol{x}(t))\boldsymbol{u}(t) \qquad (4.41)$$

where the nonlinear terms are transformed from index matrix $\boldsymbol{A}$ to $\boldsymbol{B}$. For the multivariable case, the formulation of $\boldsymbol{B}(\boldsymbol{x}(t))$ is generally not unique and it should be correctly parametrized according to the SDRE parametrization method given in [48]. Based on (4.41), State-Dependent Riccati Equation (SDRE) approach can be proposed to obtain the controllers for such nonlinear systems.

4.3.1 SDRE approach for nonlinear control design

SDRE approach mimics the standard LQR formulation that is also a systematic way of designing nonlinear feedback controllers. It is applicable to control a large class of nonlinear systems. Recent reports on SDRE approach can be found covering process control, control in robotics and power electronics application. SDRE approach offers a strong potential for improving in control of standalone and continuous tandem cold rolling process for metal strips [49]. In [50], SDRE gains with nonlinear observation and dual-SDRE-compensator gains are implemented for a high-pressure chemical vapor deposition reactor. SDRE nonlinear control is used to regulate a two-link underactuated robot constructed at UIUC and is verified on a physical plant in real time with SDRE solution repeatedly online computed [51]. For the high-speed real-time digital control of power electronics, SDRE also finds its application. SDRE-based nonlinear optimal speed controller and load torque estimator for permanent magnet synchronous motor (PMSM) are proposed and proved with the simulation and experimental results [52]. The authors in [53] apply SDRE to control a single-phase inverter, test the performance in an experimental setup and compare it with other control strategies for robustness. A comprehensive understanding of SDRE solution can be found in [54, 55, 48, 56, 57].

The SDRE optimal control problem can be formulated by finding a state feedback control law with a SDC control gain as

$$u = -K(x(t))x \tag{4.42}$$

while minimizing the same cost function as also defined for the continuous case of LQR

$$J = \frac{1}{2} \int_0^\infty x^T(t)Qx(t) + u^T(t)Ru(t) \; dt \tag{4.43}$$

The corresponding control gain with SDC can be obtained as

$$K(x(t)) = R^{-1}B^T(x(t))P(x(t)) \tag{4.44}$$

where $P(x(t))$ is the unique, symmetric and positive-definite solution for the SDRE

$$A^T(x(t))P(x(t)) + P(x(t))A(x(t)) + Q - P(x(t))B(x(t))R^{-1}B^T(x(t))P(x(t)) = 0 \tag{4.45}$$

The proof in [48] illustrates that the solution of (4.45) can locally asymptotically stabilize the compensated nonlinear system given in (4.41).

4.3.2 Solution for SDRE

In this section, one solution approach based on online gradient-type algorithm is proposed to solve the SDRE [56]. It can be regarded as a nonlinear extension of solution approach for algebraic Riccati equation of LTI systems.

Firstly, two conditions should be satisfied in order to guarantee a unique symmetric positive-definite solution $P(x)$ and a stable control law from (4.45).

1. The original SDRE (4.45) has a unique and symmetric positive-definite solution if a Cholesky factorization exists. It requires that

$$G_1(P(x), L(x)) = [g_{1,jk}] = L(x)L^T(x) - P(x) = 0 \tag{4.46}$$

where $g_{1,jk}$ $(j, k = 1, \cdots, n)$ is the (j, k)th element of the objective function G_1, and $L(x)$ is the Cholesky factor of $P(x)$.

2. (4.45) must equal zero so that it solves for a stabilized control law.

$$G_2(P(x)) = [g_{2,jk}] = A^T(x)P(x) + P(x)A(x) + Q - P(x)N(x)P(x) = 0 \tag{4.47}$$

where $N(x) = B(x)R^{-1}B^T(x)$, and $g_{2,jk}$ is the (j,k)th element of the objective function G_2.

The two conditions (4.46) and (4.47) can be combined into an integrated Lyapunov energy function as

$$E(G_1, G_2) = \frac{1}{2} \sum_{j=1}^{n} \sum_{k=1}^{n} \left[g_{1,jk}{}^2 + g_{2,jk}{}^2 \right] \tag{4.48}$$

(4.48) equals zero if and only if both $G_1\left(P(x), L(x)\right) = 0$ and $G_2\left(P(x)\right) = 0$ are satisfied. To minimize the Lyapunov function (4.48) to zero, a matrix-oriented gradient algorithm can be developed to set the update rules for $P(x)$ and $L(x)$ by changing the variables in the direction of the negative gradient of the energy function $E(G_1, G_2)$, as

$$\frac{dP(x)}{dt} = -n_P \frac{\partial E}{\partial P(x)} \tag{4.49}$$

$$\frac{dL(x)}{dt} = -n_L \frac{\partial E}{\partial L(x)} \tag{4.50}$$

where n_P and n_L are positive scaling constants, whose roles are to scale the convergence rate. Particularly, the lager n_P and n_L bring faster convergence. So the convergence of the computation process can be expedited by selecting sufficiently large values of n_P and n_L. Furthermore, in order to ensure the positive definiteness of $P(x)$ in steady state, it always requires that the convergence of $L(x)$ is faster than that of $P(x)$ by setting $n_L > n_P$.

(4.49) and (4.50) can be solved based on the analysis in [58]

$$\frac{dP(x)}{dt} = -n_P \left[A(x)\Psi_1 + \Psi_1 A^T(x) + \Psi_2 \right.$$
$$\left. -N(x)P(x)\Psi_1 - \Psi_1 P(x)N^T(x) \right] \tag{4.51}$$

$$\frac{dL(x)}{dt} = -n_L \left[\Psi_2 L(x) \right] \tag{4.52}$$

where Ψ_1 and Ψ_2 are defined respectively as

$$\Psi_1\left(P(x)\right) = A^T(x)P(x) + P(x)A(x) + Q - P(x)N(x)P(x) \tag{4.53}$$

$$\Psi_2\left(P(x), L(x)\right) = L(x)L^T(x) - P(x) \tag{4.54}$$

The final algorithm can be plotted in Fig. 4.2. This online gradient algorithm is effective to solve the SDRE. A nonlinear feedback control method based on SDRE can be employed to control the nonlinear internal energy dynamics of MMCCs. However, the obvious drawback of this approach is the high online computation, which limits its application only to some simple MMCC structures, e.g., MMCC-$\mathcal{C}_3$ and -$\mathcal{K}_{3,1}$. A simulation verification for this SDRE method based on MMCC-$\mathcal{C}_3$ will be given in the next chapter.

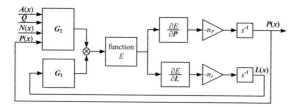

Figure 4.2: Block diagram of SDRE solution

4.4 Periodic discrete LQR

The nonlinear terms in the integrated model can also be handled by linearizing around its Steady-State Operating Point (SSOP). The remaining problem in MMCC models is the time-varying terms. In this section the LTV model is firstly modeled as a p-periodic LTV system with the chosen hyper-period, and further discretized into a periodic discrete system with the p LTI subsystems in each hyper-period (p is an integral). As a result, the LTV model can be simplified into a p-periodic discrete LTI system with a specific discretization period, which can be solved by the periodic discrete LQR (PDLQR) method. The method of periodic discrete modeling and the PDLQR design are unified and characterized by easy expansion to all the MMCC topologies. For example, the unified method is investigated and verified for MMC in [59]. Additionally, a periodic feedforward design to introduce the reference input is required to control state variables to track the references, and the periodic feedforward rejection of the measurable grid-side voltages is also needed.

4.4.1 Periodic modeling with the hyper-period

The index matrices in (3.70) for the system vector and the system input are time-varying and actually contain at least one periodic parameter. Consider the more complicated case that an MMCC connects two AC systems with different frequencies, two periodic parameters exist in the index matrices, which are the phase angles φ_1 and φ_2 with their respective periods T_1 and T_2. To model it as a periodic system, its hyper-period T_h should be determined. First write T_1 and T_2 into the irreducible fraction formats as $\frac{a_1}{b_1}$ and $\frac{a_2}{b_2}$. The hyper-period T_h can be then calculated as

$$T_h = \frac{\mathrm{lcm}(a_1,\ a_2)}{\gcd(b_1,\ b_2)} \tag{4.55}$$

where lcm(,) denotes the least common multiple of the two integers and gcd(,) calculates the greatest common divisor. For example, in the simulation example, a Hexverter connects two three-phase grids with the respective periods of $\frac{1}{50}$ s (50 Hz) and $\frac{1}{30}$ s (30 Hz). The hyper-period can be obtained as $\frac{1}{10}$ s, which also indicates that the period of the variables in the index matrices is 0.1 s.

4.4.2 Discretization of periodic LTV systems

After the hyper-period T_h is obtained, the MMCC model can be discretized by the discretization period T_d as:

$$T_d = T_h \,/\, p \tag{4.56}$$

where p is an integer number of samples per hyper-period. When the chosen value of p is large enough, the LTV system can be considered as LTI in each time interval T_d, which means the time-varying index matrices $\boldsymbol{A}(t)$ and $\boldsymbol{B}(t)$ can be treated as constant during the interval. The corresponding discrete index matrices can be written as

$$\Phi_k = e^{T_d \boldsymbol{A}(kT_d)} \tag{4.57}$$

$$\Gamma_k = \int_0^{T_d} e^{\eta \boldsymbol{A}(kT_d)} \, \boldsymbol{B}(kT_d) d\eta = (\int_0^{T_d} e^{\eta \boldsymbol{A}(kT_d)} \, d\eta) \boldsymbol{B}(kT_d) \tag{4.58}$$

The index matrices for the LTV discrete model of MMCC can numerically be calculated as

$$\boldsymbol{\Phi}_k = \boldsymbol{I} + \boldsymbol{A}(kT_d)T_d \boldsymbol{\Psi}_k \tag{4.59}$$

$$\boldsymbol{\Gamma}_k = \boldsymbol{\Psi}_k T_d \boldsymbol{B}(kT_d) \tag{4.60}$$

For example, when p is selected as 500 and the hyper-period is 0.1 s, it results a discretization time T_d of 0.2 ms. A comprehensive presentation is summarized in the periodic control theory, which can be found in [60].

The system equation (3.70) can then be discretized by the selected discretization time T_d and modeled as a periodic discrete LTI system, which is satisfied at the discretization instant k.

$$\boldsymbol{x}(k+1) = \boldsymbol{\Phi}_i \boldsymbol{x}(k) + \boldsymbol{\Gamma}_i \boldsymbol{u}(k) \tag{4.61}$$

where $\boldsymbol{\Phi}_i$ and $\boldsymbol{\Gamma}_i$ are periodic matrices, $i = \mathrm{mod}(k,p)$, $i \in \{1,\ 2,\ ,3,\ ...\ p\}$ and $k = 1,\ 2,\ 3,\ ...\ \infty$. The resultant discrete-time system (4.61) presents periodicity and time-invariant in each T_d, which can be regulated by a PDLQR with p-periodic control gains $\boldsymbol{K}_i$ ($i \in \{1,\ 2,\ ,3,\ ...\ p\}$) for each of the discretization periods.

4.4.3 Solution scheme for PDLQR

Retrieve the periodic discretized model (4.61). In each time-invariant discretization interval, the discrete-time quadratic cost function J_{PD} can be accordingly given as:

$$J_{PD} = \sum_{k=0}^{\infty} [x^T(k)Q_i x(k) + u^T(k)R_i u(k)] \qquad (4.62)$$

where Q_i and R_i ($i = \text{mod}(k, \ p)$) are the ith p-periodic error weighted matrices and the ith p-periodic control weighted matrices in the kth discretization interval, respectively. The solution method leads to the stationary algebraic Riccati equation (ARE) as

$$P_i = \Phi^T(P_i - P_i \Gamma_i (R_i + \Gamma)^{-1} \Gamma_i^T P_i) \Phi + Q_i \qquad (4.63)$$

As it is already mentioned, the solution for the above equation cannot always be analytic, but it can still be found by many numerical methods. The positive-definite solution P_i for the algebraic matrix Riccati equation (4.63) can be obtained and it will result in a control law such that

$$u(k) = -[(R_d + \Gamma_i^T P_i \Gamma_i)^{-1} P_i \Phi] x(k) = -K_i x(k) \qquad (4.64)$$

where K_i is the ith p-periodic time-invariant constant control gain matrix. The PDLQR problem is solved by determining all the p p-periodic gain matrices. [61, 62] present a clear and strict description of PDLQR design, which are based on the fundamental work [60].

A complete theory of PDLQR design includes definition of periodicity, problem formulation of discrete periodic system, factorization, lifting and standardization. The original discrete periodic Riccati equation with aperiodic solutions is step by step transformed into a stationary ARE in each discretization period by means of so-called periodic generator, so that the periodic optimal solution sequence can be obtained for this discrete periodic system. This thesis neglects the involved complex mathematical deduction for the sake of simplicity.

In a word, the PDLQR method can be understood that it solves the periodic control gains for the periodic LTI subsystems with the classic LQR method in each discretization period. The aforementioned hyper-period determination, p-periodic discrete LTI model transformation and PDLQR-based control gain design are illustrated in Fig. 4.3. To complete the whole control design, the methods for introducing a reference input and for rejecting the measurable disturbances are then proposed.

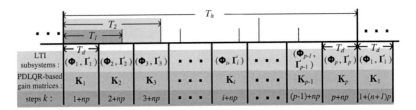

Figure 4.3: Illustration of the periodic modeling, discretization and PDLQR

4.4.4 Introduction of the reference input for periodic systems

As mentioned previously, PDLQR can simply regulate the periodic MIMO system from any initial states to the origin along the optimal trajectory defined by the cost function J_{PD}. Meanwhile, it always requires a response to a non-zero reference or a step change in actual operation. Therefore, it is necessary to introduce a reference input x_{ref} to the existing control structure to command the state variables to the desired non-zero values.

In the steady state (SS) when the reference states x_{ref} are attained, (4.61) is written as:

$$x_{ss} = x_{ref} = \Phi_i x_{ref} + \Gamma_i u_{ref,i} \tag{4.65}$$

(4.65) can be solved for

$$u_{ref,i} = \Gamma_i^{-1}(I - \Phi)x_{ref} = N_{u,i}x_{ref} \tag{4.66}$$

Note that the effect from the state feedback channel should be subtracted from the steady state control input $u_{ref,i}$. Thus the feedforward gain $N_{ff,i}$ for introducing the reference input can be designed as $N_{ff,i} = N_{u,i} + K_i$. Then the feedforward control law can be given as

$$u_{ff,i} = N_{ff,i}x_{ref} \tag{4.67}$$

where $N_{ff,i}$ are also the p-periodic constant gain matrices.

4.4.5 Feedforward rejection of disturbances for periodic systems

In order to implement disturbance rejection, a periodic model including disturbance effects is considered.

$$\begin{aligned} x(k+1) &= \Phi_i x(k) + \Gamma_i u(k) + \Gamma_{d,i} v(k) \\ &= \Phi_i x(k) + \Gamma_i(u(k) + \Gamma^{-1}\Gamma_{d,i} v(k)) \end{aligned} \tag{4.68}$$

As it is already discussed, the simplest case is when the disturbances enter the input channel and a feedforward gain $K_{d,i} = \Gamma_i^{-1}\Gamma_{d,i}$ is determined for the measurable disturbances. By applying the control component $u_{dis,i} = -K_{d,i}v(k)$ to the control input $u(k)$, disturbances in the kth discretization period can be canceled by

$$x(k+1) = \Phi x(k) + \Gamma_i(-K_{d,i}v(k)) + \Gamma_{d,i}v(k) = \Phi x(k) \qquad (4.69)$$

that is now independent on $v(k)$.

4.4.6 Complete design of PDLQR

As a conclusion, the proposed multivariable control structure is presented in Fig. 4.4. The control configuration implements optimal control for the MMCC's multiple independent current and energy variables, reference input tracking and measurable disturbance rejection. Note that all of the p-periodic matrices can be calculated and stored beforehand. By periodically applying the corresponding control gain matrix K_i and the feedforward gain matrices $N_{ff,i}$ and $K_{d,i}$ to update the control structure, the modeled MMCC system can be controlled. The presented state feedback control can be implemented by a lookup-table (LUT), which is of great advantage for real-time requirements. The complete design procedure will be implemented to MMCC-$\mathcal{K}_{3,3}$ in case study.

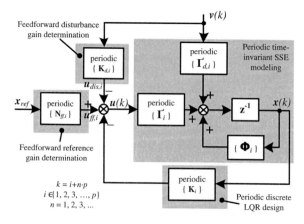

Figure 4.4: Complete PDLQR design for MMCC systems

4.5 Modulation techniques for MMCC

The modulation part is an important link between the controller and the power converter that transforms the continuous control input to the ON-OFF signals for the switchers. The modulation techniques can be generally divided into high switching frequency modulation (HSFM) and low switching frequency modulation (LSFM) according to the resulting switching frequency for the semiconductors in MMCC submodules, as illustrated in Fig. 4.5. The commonly-used HSFM methods, such as phase-shifted and level-shifted

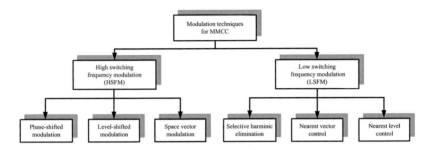

Figure 4.5: Classification of MMCC modulation techniques

modulation with triangular-wave carriers [63] and space vector modulation (SVM) [64], can be modified for MMCCs. The HSFM is well suited for conventional power converters, because it can still generate a smooth and sinusoidal current waveform disregarded for their limited voltage levels. However, for its application to MMCCs, it brings considerably high switching losses and has a restricted contribution to further improve the external current quality. Furthermore, the modulation algorithm becomes less efficient as the number of converter levels increases.

LSFM finds more applicable for MMCCs due to the efficiency for a large number of submodule semiconductors. The typical modulation methods are selective harmonic elimination (SHE), nearest vector control and nearest level control (NLC) [65]. The NLC modulation method generates the value of the voltage level by a round operation, which is simple and requires less computational efforts compare to the other methods without sacrificing MMCC performance [66].

An essential problem of MMCC modulation is to guarantee the capacitor voltage balancing among the branch submodules. The corresponding voltage balancing algorithms can be integrated into the aforementioned modulation methods, which can be categorized into three classes.

(1) Assign appropriate PWM pulses to the branch submodules without measuring their voltages. Phase-shifted modulation can inherently produce a balanced switching pattern for branch submodules, while level-shifted modulation needs certain modification to achieve the balance.

(2) Add additional closed-loop controllers for submodule voltages. The authors in [40] proposed a PI-based control strategy to implement the averaging control among the MMC branches and the balancing control among the branch submodules. A predictive controller was employed in [67] to balance the submodule voltages with respect to a predefined cost function.

(3) Apply a sorting strategy based on submodule voltages and branch current direction. The sorting strategy is frequently introduced and proves to be effective [68]. It was developed previously for MMC with cascade half-bridge submodules, but it can be adapted to other MMCCs with other submodule types. One problem associated with this strategy is the high switching frequency as the switching combination for all the submodule is updated at each control period. Instead of setting a tolerant voltage band to reduce the switching frequency, [69] proposed an improved frequency-reduced algorithm by only modifying the switching states for a small number of submodules in order to achieve the required voltage levels without the unnecessarily switching operation to the other submodules.

4.6 Summary

In this chapter multivariable control methods using optimal principle are formulated in a general description and investigated for MMCC context. Consider the LTI current dynamic submodel in the integrated model, pole placement method and LQR are proposed to control the external and circulating currents in a unified manner. To further handle the nonlinear energy submodel, nonlinear quadratic regulator is introduced. The control gain is solved based on state-dependent Riccati equation and computed online with a gradient method. The proposed SDRE-based nonlinear control requires a high online computation and cannot always ensure an analytical solution. A linearization around the steady-state operating point is applied to eliminate the model nonlinearity, and periodic discrete LQR is proposed to control the linearized integrated model. The hyper period of the both-side external systems is selected to generate the system periodicity, and periodic control gains are applied to control MMCC by searching for a lookup-table. The control performance of the optimal multivariable controllers designed for different MMCC systems will be examined in the next chapter.

5 Case studies

In order to provide a comprehensive verification for the proposed modeling and control methodology, this chapter selects three MMCC structures and their typical application scenarios as case studies. The first case study focuses on a 300kW medium-voltage DC (MVDC) transmission system based on back-to-back MMCs. The entire system is modelled into a current-energy integrated model and controlled by a cascaded PI control structure. The second case study investigates STATCOM based on MMCC-FB-$\mathcal{C}_3$, namely modular multilevel STATCOM (mmSTATCOM). It can be integrated in a medium-voltage grid compensating the reactive power to improve power quality and grid stability. A nonlinear feedback control is applied and mmSTATCOM operation under unbalanced grid-side voltages is examined. The last case study handles M3C that realizes a direct 3-phase AC/AC conversion for two different frequencies. PDLQR is chosen as the control strategy and verified for a simulation model.

5.1 MMC-based MVDC system

In the emerging trend of electric vehicle network (E-Mobility), commercial vehicles are constructed by electrified power train and auxiliaries, which is advantageous to emission reduction (CO_2, noise and exhaust gas), improved energy efficiency and better controllability. In the background that the future farms generate renewable energy locally and use it for agricultural activities, agricultural machines with innovative electrified solution can be powered by MVDC transmission system with a DC cable. It results in boost and buck voltage regulation from 800 VDC to 6 kVDC and from 6 kVDC to 750 VDC based on back-to-back MMCs.

5.1.1 System configuration and parameters

The entire MVDC system contains supply-side DC/DC converter and mobile-side DC/DC converter, as shown in Fig. 5.1. Supply-side DC/DC converter is based on a diode-clamped three-level inverter and an MMC rectifier, implementing a boost voltage conversion from 800 VDC to 6 kVDC. Mobile-side DC/DC converter is a mirrored system for supply-side one, achieving a buck conversion from 6 kVDC to 750 VDC for the tractor electric motor [70]. The system parameters are listed in Table 5.1.

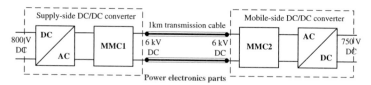

Figure 5.1: MMC-based MVDC transmission system for electric tractor

Table 5.1: System parameters of mobile-side converter

DC	DC voltage	6 kV
transmission	Cable resistance	1.15 Ω
line	Cable inductance	0.636 mH
MMC2	Rated power	300 kW
	Branch resistance	0.05 Ω
	Branch inductance	0.36 mH
	Number of branch submodules	15
	Rated submodule voltage	400 V
	Submodule capacitance	0.6 mF
	AC operating frequency	1 kHz
	Switching frequency	1 kHz
3-level rectifier	line inductance	0.2 mH × 3
	DC-side capacitance	3.3 mF × 2
	Switching frequency	5 kHz

5.1.2 Integrated modeling and cascaded PI control structure

The modeling challenges of the entire system lie on two parts, one is the back-to-back MMC configuration and another is the combined model of MMC and the three-level converter.

Model of back-to-back MMCs
As analyzed in the unified modeling method, the back-to-back MMC system can be decoupled into five parts: MMC1 AC-side equivalent circuit, MMC1 internal equivalent circuit, DC coupling link between MMC1 and MMC2, MMC2 internal equivalent circuit,

MMC2 AC-side equivalent circuit. Since MMC1 and MMC2 are mirrored systems, only the first three parts are explained for the sake of simplicity. Fig. 5.2 shows the equivalent circuits in dq frame, which are linked by the law of conservation of energy.

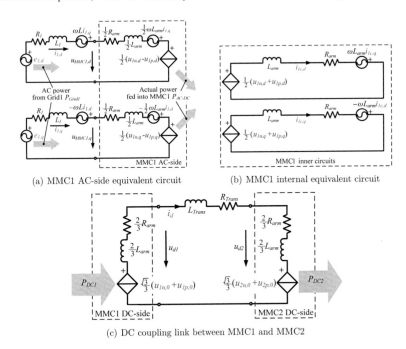

(a) MMC1 AC-side equivalent circuit (b) MMC1 internal equivalent circuit

(c) DC coupling link between MMC1 and MMC2

Figure 5.2: Equivalent circuits of back-to-back MMCs

The dq model of back-to-back MMCs under the coordinates determined by two external AC voltage sources is directly given here.

$$\dot{\boldsymbol{x}} = \boldsymbol{Ax} + \boldsymbol{Bu} + \boldsymbol{Ev} \qquad (5.1)$$

where the state variables $\boldsymbol{x} = \left(i_{1,d}\; i_{1,q}\; i_{1c,d}\; i_{1c,q}\; i_d\; i_{2c,d}\; i_{2c,q}\; i_{2,d}\; i_{2,q} \right)^T$, the dq control inputs $\boldsymbol{u} = \left(u_{1p,d}\; u_{1p,q}\; u_{1n,d}\; u_{1n,q}\; u_{0diff}\; u_{2p,d}\; u_{2p,q}\; u_{2n,d}\; u_{2n,q} \right)^T$ and AC-side disturbances $\boldsymbol{v} = \left(e_{1,d}\; e_{1,q}\; e_{2,d}\; e_{2,q} \right)^T$. Their colors correspond to Fig. 5.2. The index matrices are

$$A = \text{diag}\left(\begin{pmatrix} -\frac{R_{ac1}}{L_{ac1}} & \omega_1 \\ -\omega_1 & -\frac{R_{ac1}}{L_{ac1}} \end{pmatrix}, \begin{pmatrix} -\frac{R_{arm}}{L_{arm}} & \omega_1 \\ -\omega_1 & -\frac{R_{arm}}{L_{arm}} \end{pmatrix}, \frac{R_{dc}}{L_{dc}}, \begin{pmatrix} -\frac{R_{arm}}{L_{arm}} & \omega_2 \\ -\omega_2 & -\frac{R_{arm}}{L_{arm}} \end{pmatrix}, \begin{pmatrix} -\frac{R_{ac2}}{L_{ac2}} & \omega_2 \\ -\omega_2 & -\frac{R_{ac2}}{L_{ac1}} \end{pmatrix}\right)$$

$$B = \text{diag}\left(\begin{pmatrix} \frac{1}{2L_{ac1}} & 0 & -\frac{1}{2L_{ac1}} & 0 \\ 0 & \frac{1}{2L_{ac1}} & 0 & -\frac{1}{2L_{ac1}} \\ \frac{1}{2L_{arm}} & 0 & \frac{1}{2L_{arm}} & 0 \\ 0 & \frac{1}{2L_{arm}} & 0 & \frac{1}{2L_{arm}} \end{pmatrix}, \frac{\sqrt{3}}{3L_{dc}}, \begin{pmatrix} -\frac{1}{2L_{arm}} & 0 & -\frac{1}{2L_{arm}} & 0 \\ 0 & -\frac{1}{2L_{arm}} & 0 & -\frac{1}{2L_{arm}} \\ -\frac{1}{2L_{ac2}} & 0 & \frac{1}{2L_{ac2}} & 0 \\ 0 & -\frac{1}{2L_{ac2}} & 0 & \frac{1}{2L_{ac2}} \end{pmatrix}\right)$$

$$E = \begin{pmatrix} \frac{1}{L_{ac1}} & 0 & 0 & 0 \\ 0 & \frac{1}{L_{ac1}} & 0 & 0 \\ & 0_{5\times4} & & \\ 0 & 0 & -\frac{1}{L_{ac2}} & 0 \\ 0 & 0 & 0 & -\frac{1}{L_{ac2}} \end{pmatrix}$$

Combined model of MMC2 and three-level rectifier

The model of the mobile-side DC/DC converter is given here

$$\frac{d}{dt}\underbrace{\begin{pmatrix} i_{s1,1} \\ i_{s1,2} \\ i_{s1,3} \\ i_{s2,1} \end{pmatrix}}_{i_s} = \underbrace{\text{diag}(-\frac{1}{L_{ac}}, -\frac{1}{L_{ac}}, -\frac{1}{L_{ac}}, -\frac{R_{dc}}{L_{dc}})}_{A_t} \begin{pmatrix} i_{s1,1} \\ i_{s1,2} \\ i_{s1,3} \\ i_{s2,1} \end{pmatrix} + \underbrace{\text{diag}(-\frac{1}{L_{ac}}, -\frac{1}{L_{ac}}, -\frac{1}{L_{ac}}, -\frac{6}{L_{dc}})}_{B_t} \underbrace{\begin{pmatrix} u_{s1(1)} \\ u_{s1(2)} \\ u_{s1(3)} \\ u_{s2(1)} \end{pmatrix}}_{u_s}$$

$$-\underbrace{\frac{1}{2L_{ac}}\begin{pmatrix} \Delta u_{rec,C} & \Sigma u_{rec,C} & 0 & 0 & 0 & 0 \\ 0 & 0 & \Delta u_{rec,C} & \Sigma u_{rec,C} & 0 & 0 \\ 0 & 0 & 0 & 0 & \Delta u_{rec,C} & \Sigma u_{rec,C} \\ 0 & 0 & 0 & 0 & 0 & 0 \end{pmatrix}\begin{pmatrix} \delta_a^2 \\ \delta_b^2 \\ \delta_c^2 \\ \delta_a \\ \delta_b \\ \delta_c \end{pmatrix}}_{E_t V_s} + \begin{pmatrix} 0 \\ 0 \\ 0 \\ \frac{3}{L_{dc}}u_{dc(6kV)} \end{pmatrix}$$

$$\frac{d}{dt}\underbrace{\begin{pmatrix} i_{z,1} \\ i_{z,2} \end{pmatrix}}_{i_z} = \underbrace{\text{diag}(-\frac{R}{L}, -\frac{R}{L})}_{A_z}\begin{pmatrix} i_{z,1} \\ i_{z,2} \end{pmatrix} + \underbrace{\begin{pmatrix} \frac{1}{L} & 0 \\ 0 & \frac{1}{L} \end{pmatrix}}_{B_z}\underbrace{\begin{pmatrix} u_{z,1} \\ u_{z,2} \end{pmatrix}}_{u_z}$$

where $(i_{s1,1}\ i_{s1,2}\ i_{s1,3}\ i_{s2,1})^T$ are the three-phase AC external currents and DC external current, $(u_{s1(1)}\ u_{s1(2)}\ u_{s1(3)}\ u_{s2(1)})^T$ are the corresponding external control voltages, $(\delta_a^2\ \delta_b^2\ \delta_c^2\ \delta_a\ \delta_b\ \delta_c)^T$ are the switching states of the 3-level rectifier, $(i_{z,1}\ i_{z,2})^T$ are the circulating currents in the MMC topology and $(u_{z,1}\ u_{z,2})^T$ are the corresponding internal control voltages. R_{ac} and L_{ac} are the equivalent resistance and inductance of the AC-side external system, which are expressed as $R_{ac} = R_1 + \frac{1}{2}R$ and $L_{ac} = L_1 + \frac{1}{2}L$. R_{dc} and L_{dc} are the ones of the AC-side external system, which are $R_{dc} = 3R_2 + 2R$ and $L_{dc} = 3L_2 + 2L$. $\Delta u_{rec,C}$ and $\Sigma u_{rec,C}$ are the difference voltage and sum voltage of the two series capacitors at rectifier DC side. The IBEB model of MMCC-$\mathcal{K}_{3,2}$ and the model of capacitor voltage regulation and balancing for 3-level rectifier can be found in [59, 71], which are omitted in this case study.

Model-based control design

The model of such complicated multi-staged converter structure is usually overdetermined containing more control inputs than states. The control challenge is to realize a cooperative operation for the multiple stages. The key issue lies on the control of MMC2, which is required to receive the power from MMC1, guarantee a stable 750 V DC output and implement internal branch energy balancing.

Due to the fact that there exists no real-time communication between MMC1 and MMC2, the control task of MMC1 is to generate a stable medium voltage of 6 kV. MMC2 regards the 6 kV from MMC1 as known disturbance and regulates its DC-side input power in consideration of cable dynamics.

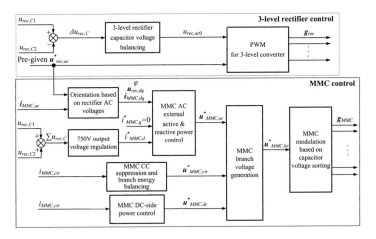

Figure 5.3: Control system for MMC-based buck converter

In the mobile-side converter, the 3-level rectifier is driven with a given phase angle and appropriately high modulation index ($m = 0.9$). The voltage balancing of the two series capacitors at rectifier DC side is realized by introducing a zero-sequence component to the modulated voltage reference for 3-level rectifier. MMC2 is controlled by taking the external AC voltages from 3-level rectifier as known disturbance. By acquiring the phase information from 3-level rectifier, the external AC current control of MMC is oriented by the external AC voltages of 3-level rectifier. A decoupled active and reactive power control is implemented for the MMC and a high power factor is achieved at the terminals of 3-level rectifier. The internal control voltage is applied to regulate the energy distribution among the branches and suppress the circulating currents in steady state. The voltage balancing task at the submodule layer is realized by the modulation method based on capacitor voltage sorting. The control block for the buck converter is briefly given in Fig. 5.3.

5.1.3 Simulation results

To examine the dynamic response for load power step and the system performance at the maximal power point, a power step from 200 kW to 300 kW is imposed in the simulation model at 0.5 s. Fig. 5.4 shows that the external and internal behaviors of MMC. It can be seen that the MMC swiftly delivers the required active power to compensate the voltage drop from the 750 V load and the internal averaging submodule voltage maintains the reference 400 V. Five-level voltage waveforms are obtained at MMC AC-side.

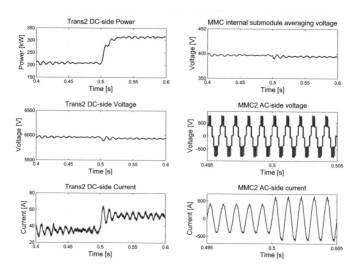

Figure 5.4: Simulation results of MMC2

The performance of branch energy balancing inside MMC2 is also examined as shown in Fig. 5.5. The six branch energy variables are initialized with different values. When the controller for IBEB is switched on, the six different branch energy start to converge to the reference. The proposed IBEB strategy can efficiently balance MMCC branch energy so that a safe and long-term operation can be achieved.

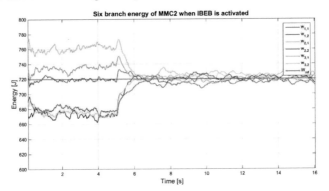

Figure 5.5: Branch energy balancing inside MMC2

5.2 MmSTATCOM for reactive power compensation

MmSTATCOM is an advanced reactive power compensator, which offers an improved transmission efficiency, a sinusoidal-shaped waveform at AC terminals, elimination of bulky and heavy harmonic filter, a flexible adaption to different voltage grades and increased converter efficiency.

5.2.1 Application scenarios and system parameters

As shown in Fig. 5.6, a three-phase grid with its voltages e_G and frequency f_1 feeds to a dynamic load through a step-down transformer. R_G and L_G are the transmission resistance and inductance. MmSTATCOM is connected to beside the load to compensate the reactive power in the grid and improve the quality of power supply. The voltage at the connection point is u_s. The mmSTATCOM is constructed by a single-delta full-bridge-configured MMCC (MMCC-C_3). The system parameters are listed in Table 5.2.

5.2.2 Operation requirement

The mmSTATCOM operation is analyzed under the following two conditions.

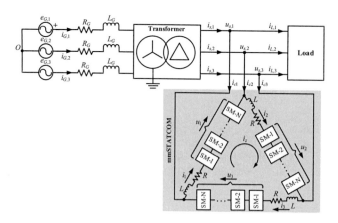

Figure 5.6: mmSTATCOM for grid reactive power compensation

Operation under balanced grid voltage condition
Under this standard operation condition, mmSTATCOM is required to compensate the dynamical reactive power generated by the loads and to guarantee a high power factor for the grid side. This task can be realized by detecting the reactive current in the grid and generating the corresponding reactive current from the converter.

Operation under unbalanced grid voltage condition
The mmSTATCOM should be controlled to have the ability of riding through severe transient voltage unbalance, and further to provide with balanced sinusoidal currents from its terminals. At the same time, the STATCOM should also be able to offer a desired amount of reactive power in order to enhance the stability of the power grid.

As described in [72], a STATCOM can operate with three different current injection targets for riding through severe transient voltage unbalance, which are balanced sinusoidal currents, constant reactive power and constant active power. In this case study balanced sinusoidal currents is required for mmSTATCOM under unbalanced grid voltages. This target aims to cancel the negative-sequence components from the external currents, which are expressed in the rotating dq frame as

$$\begin{cases} i^-_{c,d(ref)} = 0 \\ i^-_{c,q(ref)} = 0 \end{cases} \tag{5.2}$$

where $i^-_{c,d(ref)}$ and $i^-_{c,q(ref)}$ is reference values of the negative-sequence external currents of the STATCOM. It offers a balanced and sinusoidal external currents and behaves better

Table 5.2: System parameters of MMCC-FB-$\mathcal{C}_3$-based STATCOM

Power grid	Grid voltage	E_G	132 kV
	Grid frequency	f	50 Hz
	Grid resistance	R_G	3.618 Ω
	Grid inductance	L_G	100.8 mH
Transformer	Rated power	S_T	100 MVA
	Primary-side resistance	R_p	1.426 Ω
	Primary-side inductance	L_p	45.4 mH
	Secondary-side resistance	R_s	0.0462 Ω
	Secondary-side inductance	L_s	0.7353 mH
mmSTATCOM	Compensated power rating	S_{conv}	100 MVar
	External voltage	U_s	30.4 kV
	Branch resistance	R	0.6226 Ω
	Branch inductance	L	18 mH
	Number of branch submodules	N_{SM}	10
	Submodule capacitance	C_{SM}	1.9 mF
	Rated submodule voltage	$U_{SM,ref}$	9 kV
	Submodule voltage range		4.5 kV $< U_{SM} <$ 12 kV

than the other two ones by supporting positive-sequence grid voltages.

5.2.3 Generalized modeling for MMCC-$\mathcal{C}_3$

As the MMCC structure used in the investigated mmSTATCOM system is classified as MMCC-$\mathcal{C}_3$. The procedure and results obtained in Chapter 3 can be directly applied. Refer to Fig. 5.6, $u_{s,1}$, $u_{s,2}$ and $u_{s,3}$ are the external voltages of MMCC-$\mathcal{C}_3$, which are the bus voltages or PCC (Point of Common Coupling). i_{c1}, i_{c2} and i_{c3} are the external currents.

Total current dynamic model
Based on the dynamic equations for the branch current and the KCL equations at the terminals of MMCC-$\mathcal{C}_3$, the external current dynamic equations can be computed as

$$\begin{cases} \frac{d}{dt}i_{c1} = -\frac{R}{L}i_{c1} + \frac{1}{L}(u_1 - u_3) + \frac{1}{L}(2u_{s,1} - u_{s,2} - u_{s,3}) \\ \frac{d}{dt}i_{c2} = -\frac{R}{L}i_{c2} + \frac{1}{L}(u_2 - u_1) + \frac{1}{L}(2u_{s,2} - u_{s,1} - u_{s,3}) \end{cases} \qquad (5.3)$$

As the circulating current i_z in MMCC-$\mathcal{C}_3$ can be expressed as the sum of the three branch currents divided by three, the circulating current dynamic equation is

$$\frac{d}{dt}i_z = -\frac{R}{L}i_z - \frac{1}{3L}(u_1 + u_2 + u_3) \tag{5.4}$$

Internal branch energy balancing model
Refer to the IBEB model of MMCC-$\mathcal{C}_3$ derived before, the three branch energies can be regulated as

$$\frac{d}{dt}\begin{pmatrix} w_{\Sigma,\Delta} \\ w_{BG(s1),\Delta1} \\ w_{BG(s1),\Delta2} \end{pmatrix} = \begin{pmatrix} u_{s,1} - u_{s,3} & u_{s,2} - u_{s,3} & 0 \\ 0 & 0 & -\frac{3}{2}u_1 \\ 0 & 0 & -\frac{3}{2}u_2 \end{pmatrix} \begin{pmatrix} i_{c1} \\ i_{c2} \\ i_z \end{pmatrix} \tag{5.5}$$

By regulating the above three energy variables to zero, the three branch energies are controlled to the energy reference. It can be inspected that the terms $u_{s,1} - u_{s,3}$ and $u_{s,2} - u_{s,3}$ are the time-varying terms and the elements $-\frac{3}{2}u_1$ and $-\frac{3}{2}u_2$ deals with branch voltages that introduces nonlinearity to the IBEB model. Therefore, the IBEB model of MMCC contains time-varying and nonlinearity that are required to be solved in the later multivariable control design.

Integrated current-energy model
Equations (5.3), (5.4) and (5.5)) can be integrated into a standard state-space formulation in a matrix form as

$$\frac{d}{dt}\begin{pmatrix} i_{c1} \\ i_{c2} \\ i_z \\ w_{\Sigma,\Delta} \\ w_{BG(s1),\Delta1} \\ w_{BG(s1),\Delta2} \end{pmatrix} = \begin{pmatrix} -\frac{R}{L} & 0 & 0 & 0 & 0 & 0 \\ 0 & -\frac{R}{L} & 0 & 0 & 0 & 0 \\ 0 & 0 & -\frac{R}{L} & 0 & 0 & 0 \\ u_{s,1} - u_{s,2} & u_{s,2} - u_{s,3} & 0 & 0 & 0 & 0 \\ 0 & 0 & -\frac{3}{2}u_1 & 0 & 0 & 0 \\ 0 & 0 & -\frac{3}{2}u_2 & 0 & 0 & 0 \end{pmatrix} \begin{pmatrix} i_{c1} \\ i_{c2} \\ i_z \\ w_{\Sigma,\Delta} \\ w_{BG(s1),\Delta1} \\ w_{BG(s1),\Delta2} \end{pmatrix}$$

$$+ \begin{pmatrix} \frac{1}{L} & 0 & -\frac{1}{L} \\ -\frac{1}{L} & \frac{1}{L} & 0 \\ -\frac{1}{3L} & -\frac{1}{3L} & -\frac{1}{3L} \\ 0 & 0 & 0 \\ 0 & 0 & 0 \\ 0 & 0 & 0 \end{pmatrix} \begin{pmatrix} u_1 \\ u_2 \\ u_3 \end{pmatrix} + \begin{pmatrix} \frac{2}{L} & -\frac{1}{L} & -\frac{1}{L} \\ -\frac{1}{L} & \frac{2}{L} & -\frac{1}{L} \\ 0 & 0 & 0 \\ 0 & 0 & 0 \\ 0 & 0 & 0 \\ 0 & 0 & 0 \end{pmatrix} \begin{pmatrix} u_{s,1} \\ u_{s,2} \\ u_{s,3} \end{pmatrix} \tag{5.6}$$

5.2.4 Multivariable controller design

Continue with the integrated model of the mmSTATCOM, a nonlinear feedback control based on SDRE is employed. The following procedures are implemented:

Cancellation of time-varying terms
The current dynamic part in (5.6) is transformed into the dq frame so that a decoupled control effect for AC external currents can be achieved. The time-varying terms $u_{s,1} - u_{s,2}$ and $u_{s,2} - u_{s,3}$ can therefore be replaced by the constant grid voltages $u_{s,d}$ and $u_{s,q}$ that are the dq components of the PCC voltages.

Rewrite (5.6) in a state-dependent form
In order to apply the SDRE method, the integrated model (5.6) should be further rewritten from its original form into SDRE formulation by moving the terms $-\frac{3}{2}u_1 i_z$ and $-\frac{3}{2}u_2 i_z$ from the system matrix to the input matrix. The SDRE approach mentioned before is applied to control the obtained nonlinear mmSTATCOM system.

Grid-side unbalance estimation
The unbalanced grid voltages need to be estimated by prediction estimator. Suppose the voltage sources in System1 are under unbalanced conditions and only the fundamental-frequency component is considered, the grid-side voltages $\boldsymbol{u}_{s,dq} = \begin{pmatrix} u_{s,d} & u_{s,q} \end{pmatrix}^T$ under the grid-voltage-oriented rotating dq frame can be decomposed into positive-sequence components $\boldsymbol{u}_{s,dq}^P$ and negative-sequence components $\boldsymbol{u}_{s,dq}^N$. For the neutral points O and N are not connected, the effect from zero-sequence component $\boldsymbol{u}_{s,dq}^0$ is neglected. It gives as

$$\boldsymbol{u}_{s,dq} = \boldsymbol{u}_{s,dq}^P + \boldsymbol{u}_{s,dq}^N \tag{5.7}$$

$$\begin{pmatrix} u_{s,d} \\ u_{s,q} \end{pmatrix} = \begin{pmatrix} U_{s,d}^P \\ U_{s,q}^P \end{pmatrix} + \begin{pmatrix} \cos 2\omega_1 t & \sin 2\omega_1 t \\ -\sin 2\omega_1 t & \cos 2\omega_1 t \end{pmatrix} \begin{pmatrix} U_{s,d}^N \\ U_{s,q}^N \end{pmatrix} \tag{5.8}$$

where $U_{s,d}^P$, $U_{s,q}^P$, $U_{s,d}^N$ and $U_{s,q}^N$ appear as DC values. It can be seen that in rotating dq frame the positive-sequence component $\boldsymbol{u}_{s,dq}^P$ contains pure DC values and the negative-sequence component $\boldsymbol{u}_{s,dq}^N$ is identified by sinusoidal disturbances with double System1 frequency $2\omega_1$. The positive- and negative-sequence components can be respectively modeled as

$$\dot{\boldsymbol{u}}_{s,dq}^P = \boldsymbol{0} \tag{5.9}$$

$$\ddot{\boldsymbol{u}}_{s,dq}^N = -4\omega_1^2 \boldsymbol{u}_{s,dq}^N \tag{5.10}$$

(5.10) can be adapted into a standard differential form for the d component $u_{s,d}^N$ and the q component $u_{s,q}^N$ of the negative sequence component, respectively as

$$\frac{\mathrm{d}}{\mathrm{d}t} \begin{pmatrix} \dot{u}_{s,d}^N \\ u_{s,d}^N \end{pmatrix} = \begin{pmatrix} 0 & -4\omega_1^2 \\ 1 & 0 \end{pmatrix} \begin{pmatrix} \dot{u}_{s,d}^N \\ u_{s,d}^N \end{pmatrix} = \boldsymbol{M}(\omega_1) \begin{pmatrix} \dot{u}_{s,d}^N \\ u_{s,d}^N \end{pmatrix} \tag{5.11}$$

$$\frac{\mathrm{d}}{\mathrm{d}t}\begin{pmatrix}\dot{u}_{s,q}^{N}\\ u_{s,q}^{N}\end{pmatrix}=\begin{pmatrix}0 & -4\omega_{1}^{2}\\ 1 & 0\end{pmatrix}\begin{pmatrix}\dot{u}_{s,q}^{N}\\ u_{s,q}^{N}\end{pmatrix}=\boldsymbol{M}(\omega_{1})\begin{pmatrix}\dot{u}_{s,q}^{N}\\ u_{s,q}^{N}\end{pmatrix} \tag{5.12}$$

By defining the disturbance variables $\boldsymbol{w}_{d,s}=\left(u_{s,d}^{P}\ \ u_{s,q}^{P}\ \ u_{s,d}^{N}\ \ u_{s,q}^{N}\right)^{T}$ and the states

for modeling the disturbances $\boldsymbol{x}_{d,s}=\left(u_{s,d}^{P}\ \ u_{s,q}^{P}\ \ \dot{u}_{s,d}^{N}\ \ u_{s,d}^{N}\ \ \dot{u}_{s,q}^{N}\ \ u_{s,q}^{N}\right)^{T}$, its disturbance

model can be given as

$$\begin{cases}\dot{\boldsymbol{x}}_{d,s}= & \mathrm{diag}\left(\boldsymbol{0}_{2\times2},\ \boldsymbol{M}(\omega_{1}),\ \boldsymbol{M}(\omega_{1})\right)\boldsymbol{x}_{d,s}=\boldsymbol{A}_{d,s}\boldsymbol{x}_{d,s}\\[2mm] \boldsymbol{w}_{d,s}= & \begin{pmatrix}1 & 0 & 0 & 0 & 0 & 0\\ 0 & 1 & 0 & 0 & 0 & 0\\ 0 & 0 & 0 & 1 & 0 & 0\\ 0 & 0 & 0 & 0 & 0 & 1\end{pmatrix}\boldsymbol{x}_{d,s}=\boldsymbol{C}_{d,s}\boldsymbol{x}_{d,s}\end{cases} \tag{5.13}$$

A prediction estimator can be designed to estimate the unbalanced grid voltages.

5.2.5 Simulation verification

A detailed switch model of the mmSTATCOM system is established in Simulink with the help of Simpower toolbox. The following parameters are given to the controller: $\boldsymbol{Q}=\mathrm{diag}(200,\ 200,\ 18,\ 12,\ 7,\ 7)$, $\boldsymbol{R}=\mathrm{diag}(500,\ 500,\ 500)$, $n_{p}=-0.05$, $n_{L}=-0.5$. The control strategy is firstly tested with a state-space model of the mmSTATCOM system and then applied to the simulink model. The simulation verification is conducted under the following three operation scenarios:

Scenario I: Compensation of the varying reactive power
The load variation is shown in Table 5.3. As comparatively larger weights are given to limit the control input and the branch energy control, the tracking of the load reactive power becomes slightly slower than PI controller, but the proposed multivariable controller can also guarantee a close steady-state tracking for the reactive power reference. Fig. 5.8 shows the internal branch energies of the mmSTATCOM. The DC components of the averaging branch SM voltages keep stable and almost unchanged, which is apparently better than the PI controller. The sum branch energy is under control and the internal branch energy balancing is well achieved.

Scenario II: SM capacitor voltage tracking control
The tracking of the reference of the SM voltage is verified under this scenario. The mmSTATCOM is required to track a step change of SM reference voltage without any apparent distortion to the external reactive power compensation. A constant load with

Table 5.3: Setting of the varying loads

Time [s]	Active power [MW]	Reactive power [MVar]
0 - 0.04	0	0
0.04 - 0.12	50	30
0.12 - 0.24	60	60
0.24 -	60	-60

40 MW active power and 40 MVar reactive power is connected to the grid. At $t = 0.2$ s, the reference value of SM voltage raises to 10 kV. Fig. 5.9 shows the waveform of the averaging capacitor voltage of the three branches and the dq external current. The DC component of the capacitor voltage is controlled to tend towards the new reference value 10 kV with a setting time 0.4675 s, which means that after 0.4675 s the averaging capacitor voltage will stay within the 5 % range of the new reference. These three branches even exhibit a balanced energy raise during the dynamics. Although the dynamic response of the SM capacitor voltage is not quite fast, the external behavior of the mmSTATCOM to compensate the reactive power is not influenced. The active current $i_{c,d}$ shows that a certain amount of energy is absorbed from the grid in order to charge the SM capacitors, while the reactive current $i_{c,q}$ remains the same.

Scenario III: Operation under unbalanced grid voltage condition

The most common unbalanced fault, single-phase fault, is considered as the unbalanced grid voltage condition here. The amplitude of the phase-a grid voltage $v_{s,1}$ is reduced to 50 % at 0.12 s due to the temporary line-to-ground fault. During the fault, the mmSTAT-COM can still provide three-phase balanced sinusoidal external currents. Additionally, it is able to generate a desired amount of reactive power to enhance the stability of the power system. At 0.4 s, the grid fault is cleared, and the STATCOM returns to the normal operation.

Fig. 5.10 shows that the negative-sequence components in the external currents $i_{c,d}$ and $i_{c,q}$, whose double-frequency components are almost eliminated. The external currents from the mmSTATCOM are sinusoidal and balanced. The reactive power from the load is compensated even in the fault duration. A double-frequency oscillation of the load reactive power is caused by the interaction of the unbalanced three-phase voltages and currents in the grid. The positive-sequence reactive current is correctly compensated and therefore the unfaulted phases operate at high power factors. Fig. 5.11 shows the internal branch energy and the circulating current. Even under the unbalanced PCC voltages, the sum branch energy is under well control, and the three branch energies stay stable but not totally balanced. Since the fault resolves, the balance among the three branch energies is again achieved.

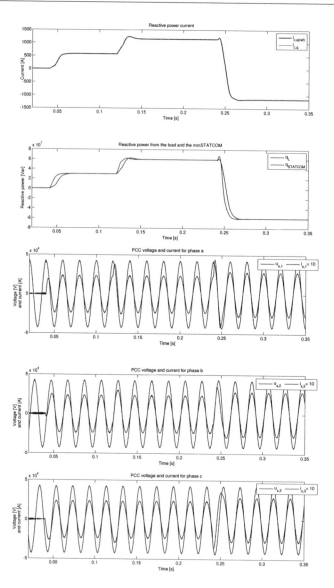

Figure 5.7: Scenario I: Grid-side performance under balanced grid voltages

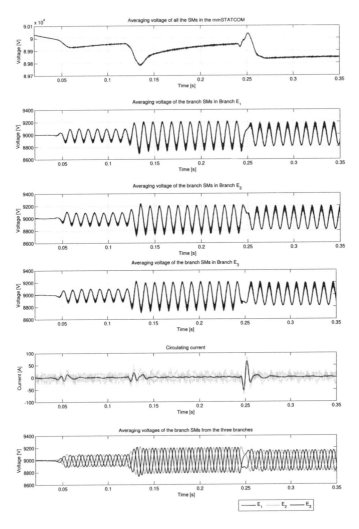

Figure 5.8: Scenario I: mmSTATCOM internal branch energy under balanced grid voltages

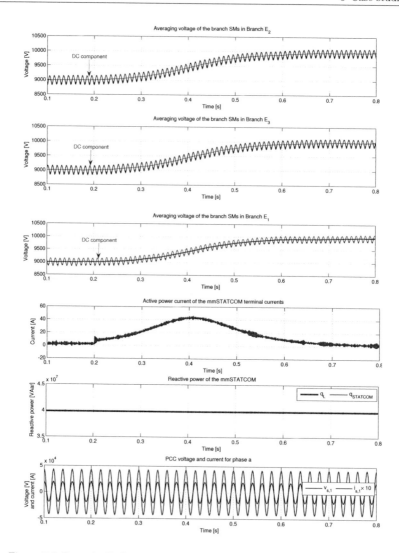

Figure 5.9: Scenario II: Step response of SM voltage under balanced grid voltages

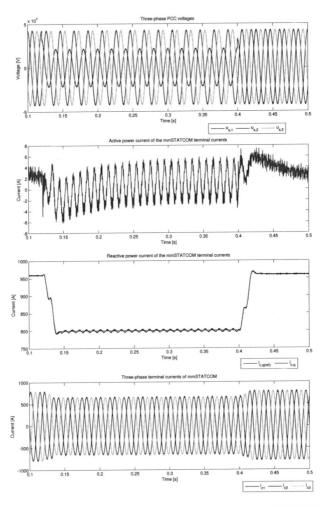

Figure 5.10: Scenario III: Voltage and current waveforms of mmSTATCOM terminals under unbalanced grid voltages

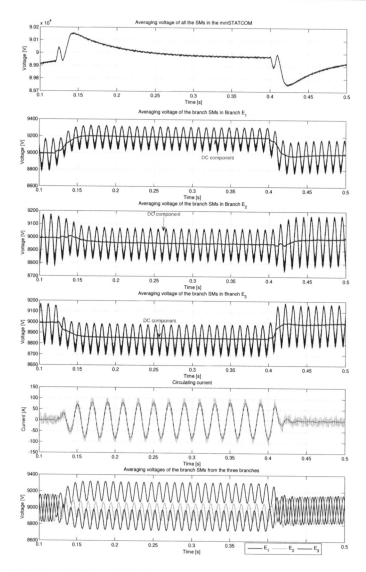

Figure 5.11: Scenario III: mmSTATCOM internal branch energy under unbalanced grid voltages

5.3 M3C for active power transmission

A direct AC/AC multilevel converter, MMCC-$\mathcal{K}_{3,3}$ or M3C, is considered here as the last case study. Compare with AC/AC conversion based on BTB-MMCs, M3C implements a direct AC/AC conversion without a DC-link. It can be applied to interconnect two different three-phase AC grids and drive electrical machines.

5.3.1 M3C-based active power transmission system

The application scenario is shown in Fig. 3.6, where the M3C structure connects two three-phase AC systems with different voltage magnitudes and different frequencies. As this M3C system is always chosen as example for model development. The procedure of CVCS-based equivalent model of M3C, the external modeling, the internal modeling and their combination of integrated current-energy model can be found in the text before. By following the matrix arrangement given in (3.70), the model of M3C in dq frames are directly given here.

The state variables including current and energy are listed here:

$$
\begin{cases}
\begin{aligned}
i_{s(dq)} &= \left(i_{s1,d}\ i_{s1,q}\ i_{s2,d}\ i_{s2,q}\right)^T \\[4pt]
i_z &= \left(i_{z1}\ i_{z2}\ i_{z3}\ i_{z4}\right)^T
\end{aligned} \\[6pt]
\begin{pmatrix}
w_\Sigma \\
w_{P(s1),\Delta1} \\
w_{P(s1),\Delta2} \\
w_{P(s2),\Delta1} \\
w_{P(s2),\Delta2} \\
w_{BG(v1,1),\Delta1} \\
w_{BG(v1,1),\Delta2} \\
w_{BG(v1,2),\Delta1} \\
w_{BG(v1,2),\Delta2}
\end{pmatrix}
=
\begin{pmatrix}
1 & 1 & 1 & 1 & 1 & 1 & 1 & 1 & 1 \\
1 & 1 & 1 & -\frac{1}{2} & -\frac{1}{2} & -\frac{1}{2} & -\frac{1}{2} & -\frac{1}{2} & -\frac{1}{2} \\
-\frac{1}{2} & -\frac{1}{2} & -\frac{1}{2} & 1 & 1 & 1 & -\frac{1}{2} & -\frac{1}{2} & -\frac{1}{2} \\
1 & -\frac{1}{2} & -\frac{1}{2} & 1 & -\frac{1}{2} & -\frac{1}{2} & 1 & -\frac{1}{2} & -\frac{1}{2} \\
-\frac{1}{2} & 1 & -\frac{1}{2} & -\frac{1}{2} & 1 & -\frac{1}{2} & -\frac{1}{2} & 1 & -\frac{1}{2} \\
1 & -\frac{1}{2} & -\frac{1}{2} & 0 & 0 & 0 & 0 & 0 & 0 \\
-\frac{1}{2} & 1 & -\frac{1}{2} & 0 & 0 & 0 & 0 & 0 & 0 \\
0 & 0 & 0 & 1 & -\frac{1}{2} & -\frac{1}{2} & 0 & 0 & 0 \\
0 & 0 & 0 & -\frac{1}{2} & 1 & -\frac{1}{2} & 0 & 0 & 0
\end{pmatrix}
\begin{pmatrix}
w_{1,1} \\
w_{1,2} \\
w_{1,3} \\
w_{2,1} \\
w_{2,2} \\
w_{2,3} \\
w_{3,1} \\
w_{3,2} \\
w_{3,3}
\end{pmatrix}
\end{cases}
$$

The external and circulating control inputs are

$$
\begin{cases}
u_{s(dq)} &= \left(u_{s1,d}\ u_{s1,q}\ u_{s2,d}\ u_{s2,q}\right)^T \\[4pt]
u_z &= \left(u_{z1}\ u_{z2}\ u_{z3}\ u_{z4}\right)^T
\end{cases}
$$

The transformation between the decoupled inputs and original branch voltages is given as (3.27). The disturbance term for external voltage sources is

$$E_t V_s = (\frac{3}{L+3L_1}e_{s1,d} \quad \frac{3}{L+3L_1}e_{s1,q} \quad -\frac{3}{L+3L_2}e_{s2,d} \quad -\frac{3}{L+3L_2}e_{s2,q})^T \qquad (5.14)$$

The index matrices for system states are

$$
\begin{cases}
A_{t(dq)} &= \mathrm{diag}(\begin{pmatrix} -\frac{R+3R_1}{L+3L_1} & \omega_1 \\ \omega_2 & -\frac{R+3R_1}{L+3L_1} \end{pmatrix}, \begin{pmatrix} -\frac{R+3R_2}{L+3L_2} & \omega_2 \\ \omega_2 & -\frac{R+3R_2}{L+3L_2} \end{pmatrix}) \\[2mm]
A_z &= \mathrm{diag}(-\frac{R}{L}, \ -\frac{R}{L}, \ -\frac{R}{L}, \ -\frac{R}{L}) \\[2mm]
A_{w\Sigma} &= (\frac{3}{2}e_{s1,d} \quad \frac{3}{2}e_{s1,q} \quad -\frac{3}{2}e_{s2,d} \quad -\frac{3}{2}e_{s2,q})^T \\[2mm]
A_{bal} &= \begin{pmatrix} -\frac{3}{2}e_{s2,1}+\frac{3}{2}e_{s2,3} & -\frac{3}{2}e_{s2,2}+\frac{3}{2}e_{s2,3} & 0 & 0 \\ 0 & 0 & -\frac{3}{2}e_{s2,1}+\frac{3}{2}e_{s2,3} & -\frac{3}{2}e_{s2,2}+\frac{3}{2}e_{s2,3} \\ -\frac{3}{2}e_{s1,1}+\frac{3}{2}e_{s1,3} & 0 & -\frac{3}{2}e_{s1,2}+\frac{3}{2}e_{s1,3} & 0 \\ 0 & -\frac{3}{2}e_{s1,1}+\frac{3}{2}e_{s1,3} & 0 & -\frac{3}{2}e_{s1,2}+\frac{3}{2}e_{s1,3} \end{pmatrix}
\end{cases}
$$

The index matrices for control input are

$$
\begin{cases}
B_{t(dq)} &= \mathrm{diag}(-\frac{1}{L+3L_1}, \ -\frac{1}{L+3L_1}, \ -\frac{1}{L+3L_2}, \ -\frac{1}{L+3L_2})^T \\[2mm]
B_z &= \mathrm{diag}(-\frac{1}{3L}, \ -\frac{1}{3L}, \ -\frac{1}{3L}, \ \frac{1}{3L})
\end{cases}
$$

The M3C is controlled in the rotating dq frames, and a pure active power transmission from System1 to System2 can be realized. The parameters of the simulated system are listed in Table 5.4.

Table 5.4: Parameters of M3C system

MMCC-$\mathcal{K}_{3,3}$	Branch resistance	R	$0.3\,\Omega$
	Branch inductance	L	$1.2\,\mathrm{mH}$
	Number of SMs per branch	N	10
	Nominal SM voltage	U_{SM}	$100\,\mathrm{V}$
	SM capacitance	C	$3.3\,\mathrm{mF}$
System1	Voltage magnitude	U_1	$220\,\mathrm{V}$
	Voltage frequency	f_1	$50\,\mathrm{Hz}$
	Resistance	R_1	$1\,\Omega$
	Inductance	L_1	$2\,\mathrm{mH}$
System2	Voltage magnitude	U_2	$110\,\mathrm{V}$
	Voltage frequency	f_2	$10\,\mathrm{Hz}$
	Resistance	R_2	$2\,\Omega$
	Inductance	L_2	$3\,\mathrm{mH}$

5.3.2 Simulation verification

The performance of the controlled M3C system is verified under the following two conditions. The first condition is to test the step response of the active current command while guarantee the reactive current zero, so that a required amount of active power is delivered from System1 to System2. The second condition verifies that the averaging capacitor voltage from all the 90 submodules can be exactly controlled and well balanced during operation.

Scenario I: Active current control

The simulation conditions can be described as: the active current $i_{s2,d}$ for System2 is firstly set to 22 A, at 0.4 s changed to 39 A and at 0.8 s changed back to 22 A. During the process, the both-side reactive currents $i_{s1,q}$ and $i_{s2,q}$ are kept to zero.

Fig. 5.12 shows the controlled active current and reactive current from the both-side systems and the corresponding transformed branch voltages in dq frames. The PDLQR gives a quick response to the reference input change and a uniform convergence rate for all the external current variables. The proposed PDLQR inherently deals with the complicated coupling among the current variables and achieves an overall good compromise between state variable tracking and control effort saving. Fig. 5.13 shows the grid voltages, grid currents and M3C external voltages from the both-side systems. The grid voltages $\boldsymbol{u}_{s1}$ are the same phase as the grid currents $\boldsymbol{i}_{s1}$, and the grid voltages $\boldsymbol{u}_{s2}$ are the opposite phase with the grid currents $\boldsymbol{i}_{s2}$. A pure active transmission for the both-side systems is achieved. It can be seen that the branch variables contain the double frequency components. After the decomposition method for the branch currents and the branch voltage transformation, the double frequency components in the branch variables can be precisely controlled under the rotating dq frames. The internal branch energy is shown in Fig. 5.12. It can be seen that all the nine branch energies are stable during the operation. Their DC offsets are filtered and plotted in the last subfigure. At each step instant the internal branch energy distribution is influenced to certain degree, but it can again return to balance after some adjustment time. The internal branch energies are controlled into a slow dynamics, because comparatively smaller weighting factors are assigned to the IBEB weighting submatrix in the total cost function than the ones in the external control submatrix.

Scenario II: Internal capacitor voltage control

The second condition tests the internal capacitor voltage control and balancing. Firstly, the capacitor voltage reference is changed from 100 V to 110 V at 0.4 s and after 0.4 s it returns to 100 V. During this process, the active power obtained at System2 is fixed and unity power factors are achieved for the both-side systems.

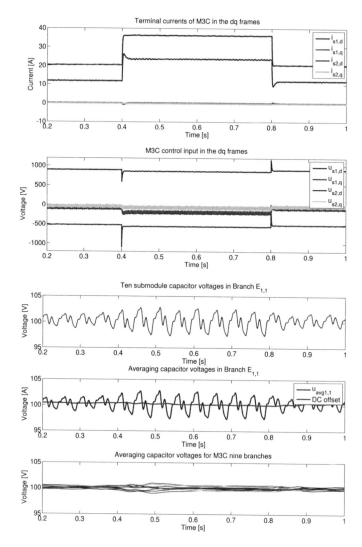

Figure 5.12: Scenario I: dq external variables and internal branch energy

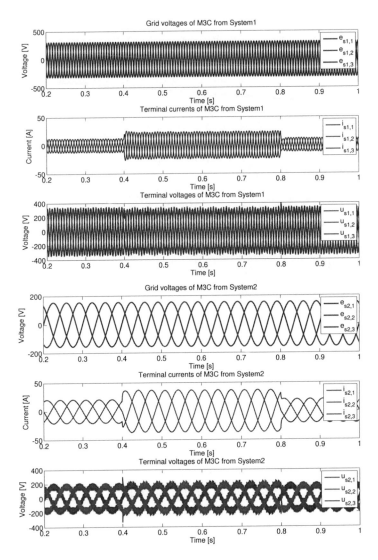

Figure 5.13: Scenario I: grid side voltages and currents in the *abc* frame

Fig. 5.14 shows the internal capacitor voltages of the M3C system. In the first subfigure the averaging capacitor voltage from all the submodules is controlled exactly at its reference after 0.2 s settling time. In the second subfigure, the nine averaging capacitor voltages from the nine branches are balanced even though they have certain differences during the transient.

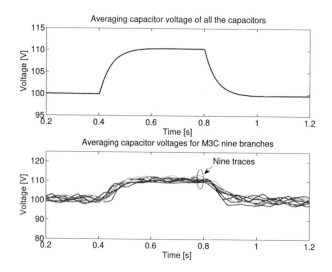

Figure 5.14: Scenario II: M3C averaging capacitor voltages from all the submodules and from each branch

Fig. 5.15 shows the external currents in the dq frames and in the abc frame. At 0.4 s, an additional amount of active power is extracted from System1 and stored in the submodule capacitors in the M3C. The proposed decoupled control strategy can allow regulation of capacitor voltage without influencing the external current behavior at System2 side. At 0.8 s, the capacitor voltages are required to be discharged to 100 V. The energy stored in the M3C capacitors is used to support the active power transmission to System2 and the additional amount is fed back to System1. It can be observed from the first subfigure in Fig. 5.15 that $i_{s1,d}$ is negative during the transient, because a certain amount of the capacitor energy flows to System1 for a fast reference tracking of capacitor voltages.

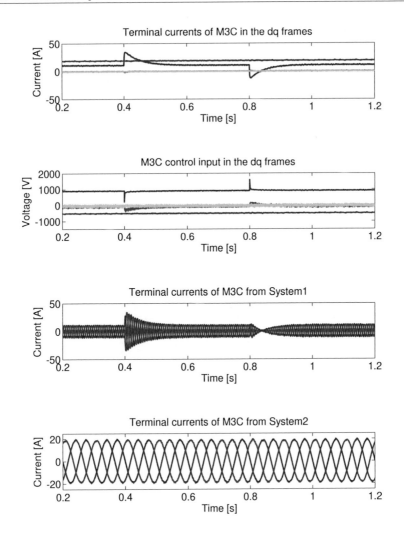

Figure 5.15: Scenario II: M3C external currents in the dq frames and in the abc frame

5.4 Summary

In this chapter three different MMCC-based systems are chosen as examples to illustrate the proposed modeling procedure and multivariable control methods. The external performance of generating specific external currents as well as required power components is realized. Meanwhile, the internal branch energy balancing and submodule voltage regulation is guaranteed during the normal operation. Furthermore, a down-scaled prototype based on classic MMC is developed in laboratory and is used to verify the modeling and control concepts.

6 Hardware setup of down-scaled MMCC system

To validate the proposed modeling and control methods throughout the work a prototype of MMCC-$\mathcal{K}_{3,2}$, e.g., three-phase MMC, is developed. The schematic diagram of the entire MMCC system is presented in Fig. 6.1. Fig. 6.2 shows a photo of the corresponding hardware setup. The key structure is the three-phase MMC with a rating power of 1 kW and 5 half-bridge submodules per branch, which can flexibly be configured as a DC/AC inverter or a AC/DC rectifier according to the requirement. In this thesis the DC/AC inversion operation is focused for the developed MMC system, which connects to a 200 V DC power supply and feeds to a diode-clamped three-level rectifier and resistive load.

6.1 Overview of buck converter

The entire hardware setup of MMCC system can generally be divided into two parts: power part and electronic part. To avoid the high potential from the power part, a strict galvanic isolation based on optocouplers is realized between the both parts.

The power part is composed of the developed three-phase MMC structure (MMCC-$\mathcal{K}_{3,2}$), a measurement board, two bipolar power supplies (Model BOP 100-4M), three-level rectifier and the resistive load. The three-phase MMC structure is constructed by six converter branches in double-star configuration, where each branch contains five cascaded identical half-bridge submodules and one branch inductor. That is to say, totally 30 submodules exist in this MMC. The measurement board has the power connectors for the six converter branches as well as for the three-phase load, meanwhile it also contains the signal outputs for the measured branch currents and AC external voltages. Furthermore, the measurement board is equipped with six comparators, which generates the branch current direction for implementing the classic sorting-based modulation strategy in the FPGA board. The conversion task of the MMC structure is to change the existing DC energy from the two series-connected power supplies to AC energy for feeding the load. The 3-level rectifier and the resistor are treated as load here.

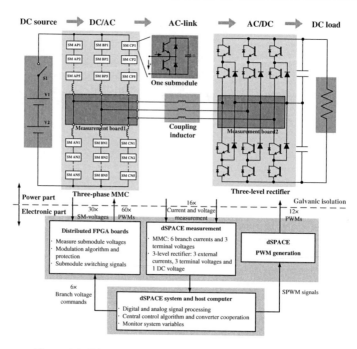

Figure 6.1: Schematic diagram of MMC-based buck converter

The electronic part adopts a layered structure, composed of central controller, measurement board, branch FPGA control boards and PWM board for three-level converter. The central controller consists of one dSPACE system and a host PC, that collects the measured signals from the measurement board to make external control decision and then generates the analog signals of the six branch voltage references. The host PC is connected to the dSPACE system that monitors the system state variables and can dynamically adjust its operating states. Due to the limited number of pins and comparatively slow control frequency, a FPGA board with abundant user-configurable IOs is used to process the branch voltage references, detect the branch submodule voltages and the branch current polarities and finally generate the PWM signals for each specific submodule. One FPGA board is in charge of one MMC phase, e.g., two converter branches. The PWM generation board for three-level converter is an extension board of dSPACE, which offers enough PWM channels, communicates with central controller and generates the driving signals for three-level converter. Details of the power and electronic parts of the hardware setup will be explained in the following sections.

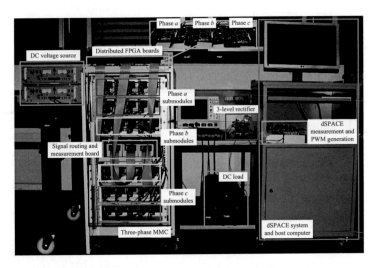

Figure 6.2: Hardware setup of MMC-based buck converter

6.2 Three-phase MMC structure

A detailed schematic diagram of the entire MMC system is shown in Fig. 6.3, whereby the power part is on the right side. The parameters of the developed three-phase MMC system are listed in Table 6.1. To distinguish each of the internal submodules, they are named as SMijk, where $i = A, B, C$ indicating the phase sequence, $j = p, n$ representing positive or negative branch and $k = 1, 2, ..., 5$ showing the kth submodule in one branch.

To initialize the MMC system all the submodule are firstly set to Block mode so that all the submodule capacitors are isolated from the external voltage sources. The switch $S1$ and the warm-up DC source $V1$ are applied to precharge the submodules one by one to the given voltage reference. To enable a balanced precharging operation, only one submodule in each phase is switched on while the rest are switched off, so that $V1$ can charge its capacitor voltage through a charging resistor. The DC sources $V2$ and $V3$ and the switch $S2$ are put into function for the inverter test. The switch $S3$ is switched on for a direct connection between the middle point of the series DC sources and the neutral point of the three-phase load. The fundamental unit of the hardware setup is the half-bridge submodule, which is firstly explained.

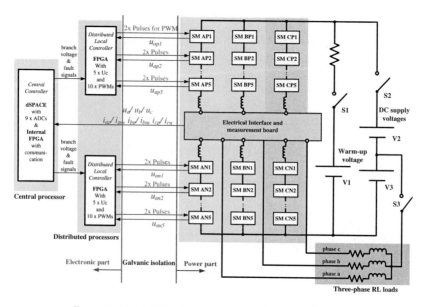

Figure 6.3: Detailed setup of the three-phase MMC system

Table 6.1: Technical data of the implemented MMC system

Norminal power	$P_{MMC,N}$	1 kW
DC-side voltage	$u_{dc,N}$	200 V
Submodule nominal voltage	$U_{SM,N}$	50 V
Submodule capacitance	$C_{SM,avg}$	2.3 mF
Number of branch submodules	N_{SM}	5
Dead time for submodule switches	T_{dt}	1.6 μs
Branch inductor	L	1 mH
Branch resistor	R	0.1 Ω
Maximal branch current	$I_{br,N}$	15 A
Load inductance	L_{AC}	2 mH
Load resistance	R_{AC}	4.5 Ω
System sampling frequency	T_s	10 kHz
Submodule switching frequency	f_{sw}	1 kHz

6.2.1 MMC submodule

Fig. 6.4(a) illustrates the circuit diagram of the developed submodule board, and Fig. 6.4(b) shows the hardware of one half-bridge submodule board, which is designed in Eurocard size.

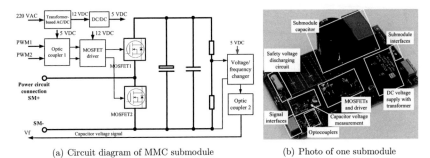

(a) Circuit diagram of MMC submodule   (b) Photo of one submodule

Figure 6.4: One submodule in the developed MMC system

Isolated power supply

A cascaded connection of MMC submodules leads to a situation that the negative connector "SM−" of each submodule has its own floating potential. To protect the electronic components on the submodule board as well as the digital processors, each submodule has to be powered by a galvanically isolated voltage source. On one submodule board MYRRA AC/DC switch power block integrated with a transformer is selected to produce a potential-free +12 V DC for the MOSFET driver. The other digital ICs are chosen with standard +5 V logic CMOS technology so that less power loss and a higher efficiency can be expected. A linear regulator is then employed to generate a stable and smooth output voltage of +5 V.

Optocoupler for circuit isolation

Since all the submodule boards have their own potentials and the digital signals from the FPGA board has to connect to them all, a circuit isolation in between is necessary. The optocoupler ACPL-064L from Avago that guarantees a high transmission speed with 12 ns turn-on time and 10 ns turn-off time is selected to realize the isolation by means of a light LED. Due to the fact that certain forward current is required to drive the light LED on, the signal from FPGA should be amplified to a current driving ability of 10 mA by a Schmitt trigger inverter (SN74LVC2G14).

Power MOSFET and gate driver chip

Power MOSFET is the key component in the submodule board and in the power electronics system. According to the hardware specification, MOSFET U3410PbF from International Rectifier with $V_{DSS} = 100$ V, $R_{DS(on)} = 0.1$ Ω and $I_D = 17$ A is applied for the submoduel board. The maximal switching frequency for one MMC submodule is designed for 10 kHz, so the turn-on and turn-off time is approximated as 2 % of the switching period, which is 2 μs. After checking the MOSFET datasheet, the MOSFET gate voltage must guarantee at least 30 mA current driving ability. Due to the limited drive ability from the optocoupler output and requirement of a level-shifted gate voltage for the high-side MOSFET, a MOSFET driver AUIRS2191S also from IR with bootstrap technique is employed, which is cheap in price and also simple for application. A bootstrap diode (ultra-fast IN4148) and a bootstrap capacitor (ceramic capacitor of 220 nF) are the only two additional passive components to construct the bootstrap driver circuit. One critical problem of bootstrap driver is that the level-shifted gate voltage for the high-side MOSFET is generated from the bootstrap capacitor and the bootstrap capacitor can only be charged through a closed-loop circuit constituted by the switched-on low-side MOSFET. This problem must be considered and avoided during the hardware debug, MMC startup and normal operation. Additionally, since the selected MOSFET driver cannot ensure in hardware a dead time of the switching between the upper and lower MOSFET, a dead time of 1.6 μs is programmed in the FPGA board for our specific application.

Submodule voltage measurement
The submodule capacitor voltages are to be monitored at each time instant, so that the modulator or the direct controller can provide a stable switching sequence to balance the submodule voltages. A voltage-frequency converter (VFC) AD7740 from Analog Devices is used to modulate the down-scaled capacitor voltage to a digital frequency signal with a output frequency from 100 kHz to 900 kHz. The down-scaled capacitor voltage is obtained by a voltage divider based on two series-connected precise resistors with 0.1 % accuracy and is filtered by a 13 kHz low-pass filter. The modulated frequency signal is then fed to the FPGA board with a clock frequency of 50 MHz, and each impulse can exactly be detected by the FPGA board. Then the FPGA chip counts the number of the output pulses in a fixed time interval and computes the corresponding capacitor voltage.

The VFC chip maintains a good linearity through the whole frequency range as shown in Fig. 6.5. By counting the frequency signal f_{VFC} (kHz) from each submodule, its capacitor voltage u_{SM} (V) can then be calculated by the following equation:

$$u_{SM} = 0.081 f_{VFC} - 8.918$$

Submodule capacitor
A paralleled structure of two capacitors are chosen as the submodule capacitors, which one electrolytic capacitor (120 V and 2.2 mF) and one film capacitor (60 V and 100 μF).

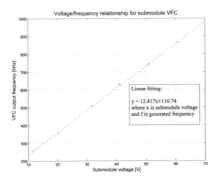

Figure 6.5: Output characteristics of voltage-frequency converter on submodule board

The film capacitor is to supply fast transient current when the submodule is switched on. For safety consideration, when the power supply for the submodule is off, the charged submodule capacitor will be automatically discharged by a normally-closed PhotoMos-Relay (AQY21SDP) in about 6 seconds.

6.2.2 Measurement board

The functional overview of the measurement board is shown in Fig. 6.6. The major components on the measurement board are six current transducers (CT, LEM LTSR25-NP) and three voltage transducers (VT, LEM LV 25-P), measuring the six branch currents and three phase-to-phase voltages. Totally nine signals generated by these transducers will be transferred to the dSPCE AD interfaces. It should be pointed out that all the required external currents and the circulating currents can be computed by these six branch currents. On the other hand, digital comparators with hysteresis reference design are implemented to obtain the polarity of the branch currents. These current polarities are also galvanically isolated by octocouplers and fed to the FPGA board.

6.3 Digital processor system

Refer to the detailed figure in Fig. 6.3, a layered digital processor system is also established. The upper-layered control structure is composed of a host PC and a dSPACE system, while the lower-layered control structure consists of modular FPGA control boards and dSPACE extension boards.

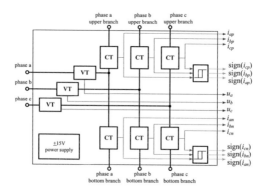

Figure 6.6: Circuit diagram of measurement board

6.3.1 Upper-layered control structure

The host computer offers a convenient interface to online configure the parameters of
the controlled MMC system and a graphic illustration of the system state variables.
With the help of the toolbox in Matlab/Simulink, the designed control algorithm can
efficiently be realized and tuned for the three-phase MMC system.

The employed dSPACE system includes the main board (DS1006) and a ADC extension
board (DS2004 A/D board). The ADC extension board collects the branch current
variables in real time and the phase-to-phase voltages and sends to the dSPACE main
board. The dSPACE main board executes the central code of the proposed algorithm and
generates the corresponding branch voltages through its analog interfaces to guarantee
an external control performance. Since the AC external voltages as well as the six branch
currents are detected in real time, the protection of the hardware in cases of over-currents
and over-voltages is ensured by setting all the generated branch voltages to zero.

6.3.2 Lower-layered control structure

The most challenging issue of driving a three-phase MMC is requirement of numerous
user-configurable IOs for measuring the submodule voltages and sending gate signals.
Since one standalone dSPACE cannot afford so many IOs, modular distributed FPGA
control boards are applied to extend the IO resource, share the computation burden for
dSPACE and boost the control frequency. The maximal sampling frequency of dSPACE
AD interfaces is only 1 kHz in "single conversion" mode, and the updating frequency

of its analog outputs is 2 kHz. With the help of FPGA parallel computing ability, the control frequency for each submodule can remarkably be raised.

The core element of the FPGA control board is XC3S700AN from Xilinx, which implements the following tasks: receive the branch voltage references from dSPACE, count the VFC pulses to determine submodule voltages, sorting for submodule voltages, multilevel modulation algorithm and switching decision based the sorting results. It also includes the functions of initialization, modulation enable, emergency block and safety measures. A flowchart for the central modulation algorithm is given in Fig. 6.7. The sorting method is based on classic bubble sort with its worst case performance $O(n^2)$ and best case performance $O(n)$.

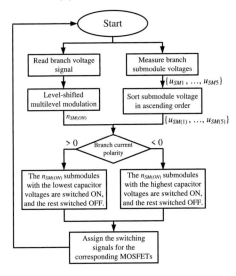

Figure 6.7: Flowchart of the implemented modulation algorithm in FPGA

6.4 Experimental results

Experiments based on the developed hardware system are carried out to verify the proposed modeling method and LQR design. The DC-side external voltage is given as 50 V, and each submodule voltage is controlled to 10 V with a switching frequency 8 kHz. One RL load is connected to the AC-side with resistance 14.38 Ω and inductance around 6 mH.

6.4.1 Steady-state operation

Fig. 6.8 shows MMC external and internal variables under the external current reference 1 A. It can be seen in Fig. 6.8(a) that the external voltage contains seven voltage levels, and the corresponding external current is smooth and sinusoidal with an amplitude of 1 A. Based on the analysis before, the value of each voltage step at the external is equal to half of the submodule voltage, e.g., 5 V. Fig. 6.8(b) shows the modulated staircase waveforms of the two branch voltages from the same phase. The branch voltage reference is a continuous signal that contains the external current control component and circulating current control component. Since the load requires only 1 A, the external voltage has seven voltage levels, and thus only four voltage levels can be observed in branch voltages. When the external reference increases, more levels of branch voltage will be put into service. Additionally, each level in the branch voltage is not constant, because the corresponding submodule voltage vary around the reference 10 V during the operation. In Fig. 6.8(c), with the help of branch energy balancing and voltage-sorting-based MMC modulation, the averaging capacitor voltage of each submodule keeps at 10 V. Due to the charging and discharging effects from branch current, the submodule capacitor has a voltage ripple of maximal 1 V. By increasing the submodule capacitance or line frequency, reducing the external current or suppressing circulating current, these voltage ripples can effectively be damped. In Fig. 6.8(d) the circulating current is suppressed to maximal 1 A in transient by regulating the circulating control component. During the experimental test, if the circulating current is not well damped, the MMC three-phase operation encounters frequently overcurrent warnings due to the resulting high branch current, which would affect the normal operation, reduce the system efficiency and even destroy submodules.

Fig. 6.9 presents the same waveforms when the external current reference is set to 1.5 A. Since the load current increases, the external voltage requires a larger amplitude, so that all the possible nine voltage levels are applied to synthesis the external voltage. The submodule voltages inside the MMC are well balanced, which can be seen from the three measured submodule voltages in Fig. 6.9(c). Although the load current increases to 1.5 A, the maximal circulating current is still limited to 1 A due to the effect of circulating current suppression.

6.4.2 Step response

The dynamic response of the controlled MMC is also examined, whereby the external three-phase currents are controlled from 1 A to 2 A as shown in Fig. 6.10. By adjusting the control parameters, different dynamic control performance can be realized.

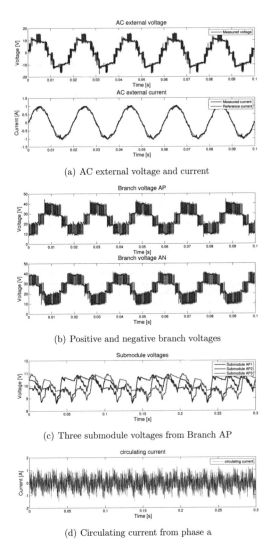

(a) AC external voltage and current

(b) Positive and negative branch voltages

(c) Three submodule voltages from Branch AP

(d) Circulating current from phase a

Figure 6.8: Experiment results for external current control at 1 A

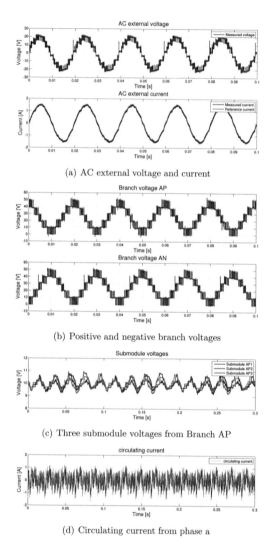

(a) AC external voltage and current

(b) Positive and negative branch voltages

(c) Three submodule voltages from Branch AP

(d) Circulating current from phase a

Figure 6.9: Experiment results for external current control at 1.5 A

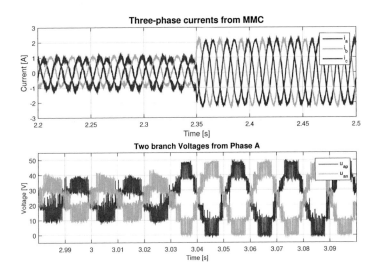

Figure 6.10: Step response of external current from 1 A to 2 A

6.5 Summary

This chapter presents the hardware configuration of the developed MMC prototype and the experimental waveforms. A three-phase MMC structure with 1 kW rating power and five submodules per branch has been successfully established in the laboratory environment and placed in an electrical enclosure. In correspondence with the modular construction of MMC, the digital control system is also implemented in a modular way, whereby a dSPACE system serves as a central control unit and three pieces of FPGA boards are in charge of the three converter phases. There are still a high number of available user-configurable IOs in the FPGA boards, which allows a future expansion for a three-phase MMC structure with a higher number of branch submodules. In the end the experimental waveforms are presented to illustrate the effectiveness of the proposed modeling and LQR methods.

Part II

MMCC direct control method

7 Modeling and formulation of direct control of MMCCs

The novel multilevel converter family named Modular Multilevel Cascaded Converter (MMCC) has become an attractive multilevel solution for medium and high voltage applications. Compared with the conventional multilevel converters, MMCC offers a large amount of voltage levels with linear expanding hardware complexity, less harmonic injection, a modular construction for redundancy, flexible scalability for different voltage requirements and a possibly reduced switching frequency as well as power losses. An MMCC structure is composed of a cascaded connection of switching submodules and is configured into a specific structural arrangement. This chapter aims to develop direct switched model, which extends the integrated model (3.70) by using submodule switching states as system inputs and adding submodule voltages as system states. After that, the problem of MMCC direct control is formulated and the high number of switching redundancy is presented in Section 7.2.

7.1 Direct switched model of MMCC

7.1.1 Model development

In order for a better illustration, the one-phase MMC (MMCC-$\mathcal{K}_{2,1}$) is employed to clarify the modeling and direct control design, as shown in Fig. 7.1. In Fig. 7.1 the two separated DC voltage sources $\frac{u_{dc}}{2}$ are given by two power amplifiers with neglectable inner resistance $\frac{R_{dc}}{2}$ and inductance $\frac{L_{dc}}{2}$. The converter branch is composed of five cascaded half-bridge submodules ($N_{SM} = 5$), denoted as HB-SMi ($i = 1, 2, ..., 5$) for Positive Branch AP and HB-SMj ($j = 6, 7, ..., 10$) for Negative Branch AN. The switching state of ith submodule from Branch AP is symboled as $g_{ap,i}$ ($g_{ap,i} \in \{0, 1\}$) and the submodule voltage is $u_{ap,i}$. The switching state and submodule voltage of the submodule from the negative branch are $g_{an,i}$ and $u_{an,i}$. The inductance of the branch inductor is denoted as L and the resistance R represents the losses from the converter branch, such as the conduction resistance $R_{ds(on)}$ of semiconductor, capacitor equivalent series resistance R_{esr} and other parasite resistance. The series inductance L_{ac} and resistance R_{ac} constitute the load of the one-phase MMC.

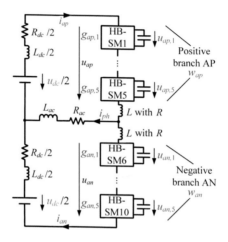

Figure 7.1: Circuit diagram of one-phase MMC (MMCC-$\mathcal{K}_{2,1}$)

The direct switched model of MMCC will be developed based on the following three parts: (1) System layer for external power control, (2) Branch layer for internal branch energy balancing and (3) Branch submodule layer for capacitor voltage balancing.

System layer for external power control

In the system layer the treatment of continuous equivalent model is still valid. As discussed before, the branch current from Branch AP and AN can be expressed by the external current and the circulating current as

$$\begin{cases} i_{ap} &= \frac{1}{2}i_{ph} + i_{cir} \\ i_{an} &= -\frac{1}{2}i_{ph} + i_{cir} \end{cases} \tag{7.1}$$

Then the differential equations for the external current and the circulating current can be obtained by applying Kirchhoff equations manually.

$$\begin{cases} \frac{d}{dt}i_{ph} &= -\frac{R+2R_{ac}}{L+2L_{ac}}i_{ph} + \frac{1}{L+2L_{ac}}\left(u_{an} - u_{ap}\right) \\ \frac{d}{dt}i_{cir} &= -\frac{R}{L}i_{cir} - \frac{1}{2L}\left(u_{ap} + u_{an}\right) + \frac{1}{2L}u_{dc} \end{cases} \tag{7.2}$$

For the direct switched model the submodule switching states and submodule capacitor voltages will be selected as the system inputs and states. Therefore, the involved branch voltages in (7.2) can be expressed by the instantaneous submodule voltages ($u_{ap,i}$ and $u_{an,i}$, $i = 1, 2, .., 5$) and the submodule switching states ($g_{ap,i}$ and $g_{an,i}$, $i = 1, 2, .., 5$) as

$$\begin{cases} u_{ap} &= \sum_{i=1}^{5} g_{ap,i} u_{ap,i} \\ u_{an} &= \sum_{i=1}^{5} g_{an,i} u_{an,i} \end{cases} \tag{7.3}$$

Branch layer for internal branch energy balancing

In this layer the branch energy dynamics should be considered for a steady long-term operation of MMCC. The instantaneous branch power of the two branches can be given as

$$\begin{cases} \frac{d}{dt} w_{ap} &= p_{ap} = i_{ap} u_{ap} = (\frac{1}{2} i_{ph} + i_{cir}) u_{ap} \\ \frac{d}{dt} w_{an} &= p_{an} = i_{an} u_{an} = (-\frac{1}{2} i_{ph} + i_{cir}) u_{an} \end{cases} \tag{7.4}$$

According to the proposed IBEB strategy, the two branch energy variables are evaluated by two equivalent ones, which are the sum energy w_Σ (for Subgoal1) and the energy difference w_Δ (for Subgoal2). Their dynamic models are given as

$$\begin{cases} \frac{d}{dt} w_\Sigma &= \frac{1}{2} i_{ph}(u_{ap} - u_{an}) + i_{cir}(u_{ap} + u_{an}) \\ \frac{d}{dt} w_\Delta &= \frac{1}{2} i_{ph}(u_{ap} + u_{an}) + i_{cir}(u_{ap} - u_{an}) \end{cases} \tag{7.5}$$

The above two dynamic equations are used to control the total energy stored in all the submodules and regulate the branch energy difference to zero. Since the AC-side current is always required to be tracked with its reference in inversion mode, the circulating current offers the flexibility to achieve the branch-layered energy balancing. The first equation in (7.5) shows that the DC component in i_{cir} combined with the sum branch voltage contributes a DC power component to change the sum branch energy. In the second equation the AC component in i_{cir} is controlled with the same frequency as the AC side, so that a DC power can be generated to compensate the branch energy difference.

Branch submodule layer for capacitor voltage balancing

This layer deals with switching the branch submodules in order to generate the required branch voltage and achieve a balanced energy distribution among submodules. This layer is usually accounted for by a submodule voltage sorting algorithm without requiring an explicit model.

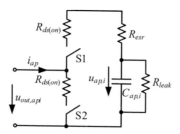

Figure 7.2: Equivalent model of one submodule

Fig. 7.2 shows an equivalent circuit of one submodule HB-SMi in Branch AP. The switches are modelled as idea switches S_1 and S_2 with their turn-on resistance $R_{ds(on)}$. The submodule switching state $g_{ap,i}$ is expressed as

$$g_{ap,i} = \begin{cases} 1 & \text{when } S1 \text{ is on \& } S2 \text{ is off} \\ 0 & \text{when } S1 \text{ is off \& } S2 \text{ is on} \end{cases}$$

As at any time instant only one switch is switched on, $R_{ds(on)}$ in the submodule can be extracted and moved out of the submodule. An additional resistance R_{ds} can be included in the branch resistance R to indicate the losses of branch submodules, which is $R_{ds} = N_{SM}R_{ds(on)}$.

The submodule capacitor is regarded as an ideal capacitor with the inevitable Equivalent Series Resistance R_{esr} and the parallel resistance R_{leak} discharging the capacitor over time. R_{esr} influences the generated voltage between submodule terminals and therefore the final branch voltage, which can also be handled by including in the branch resistance R. R_{leak} has a impact on the charging dynamics of the submodule capacitor when the submodule is switched on and discharges the capacitor when the submodule is off, which is given as

$$\begin{cases} \frac{d}{dt}u_{ap,i} = -\frac{1}{R_{leak}C_{ap,i}}u_{ap,i} + \frac{1}{C_{ap,i}}i_{ap} & \text{when } g_{ap,i} = 1 \\ \frac{d}{dt}u_{ap,i} = -\frac{1}{R_{leak}C_{ap,i}}u_{ap,i} & \text{when } g_{ap,i} = 0 \end{cases} \qquad (7.6)$$

The above two expressions can be combined as one.

$$\frac{d}{dt}u_{ap,i} = -\frac{1}{R_{leak}C_{ap,i}}u_{ap,i} + \frac{1}{C_{ap,i}}g_{ap,i}i_{ap} \qquad (7.7)$$

In the end, the dynamic model of each submodule in an MMC branch can accordingly be given as

$$\begin{cases} \frac{d}{dt}u_{ap,i} &= -\frac{1}{R_{leak}C_{ap,i}}u_{ap,i} + \frac{1}{C_{ap,i}}g_{ap,i}i_{ap} \\ \frac{d}{dt}u_{an,i} &= -\frac{1}{R_{leak}C_{an,i}}u_{an,i} + \frac{1}{C_{an,i}}g_{an,i}i_{an} \end{cases} \tag{7.8}$$

Model synthesis and feature analysis

The switched dynamic model of the one-phase MMC can be obtained by including (7.2), (7.3), (7.5) and (7.8), which is

$$\frac{d}{dt}\boldsymbol{x} = \boldsymbol{Ax} + \boldsymbol{Bu} + \boldsymbol{v} \tag{7.9}$$

where the state variables are $\boldsymbol{x} = \begin{pmatrix} i_{ph} & i_{cir} & w_{\Sigma} & w_{\Delta} & u_{ap,1} & \dots & u_{ap,5} & u_{an,1} & \dots & u_{an,5} \end{pmatrix}^T$, the control input $\boldsymbol{u} = \begin{pmatrix} g_{ap,1} & \dots & g_{ap,5} & g_{an,1} & \dots & g_{an,5} \end{pmatrix}^T$ and the disturbance term is $\boldsymbol{v} = \begin{pmatrix} 0 & \frac{1}{2L}u_{dc} & \boldsymbol{0}_{1\times 12} \end{pmatrix}^T$. The state index matrix is $\boldsymbol{A} = \begin{pmatrix} \boldsymbol{A}_{tz} & \boldsymbol{0}_{2\times 2} & \boldsymbol{0}_{2\times 10} \\ \boldsymbol{0}_{2\times 2} & \boldsymbol{0}_{2\times 2} & \boldsymbol{0}_{2\times 10} \\ \boldsymbol{0}_{10\times 2} & \boldsymbol{0}_{12\times 2} & \boldsymbol{A}_{sm} \end{pmatrix}$ with

$\boldsymbol{A}_{tz} = \begin{pmatrix} -\frac{R+2R_{ac}}{L+2L_{ac}} & 0 \\ 0 & -\frac{R}{L} \end{pmatrix}$ and $\boldsymbol{A}_{sm} = \mathrm{diag}\left(-\frac{1}{R_{leak}C_{ap,1}}, \dots, -\frac{1}{R_{leak}C_{an,5}}\right)$. The control

index matrix is $\boldsymbol{B} = \begin{pmatrix} \boldsymbol{B}_{11} \\ \boldsymbol{B}_{21} \end{pmatrix}$, $\boldsymbol{B}_{11} = \begin{pmatrix} -\frac{u_{ap,1}}{L+2L_{ac}} & \dots & -\frac{u_{ap,5}}{L+2L_{ac}} & \frac{u_{an,1}}{L+2L_{ac}} & \dots & \frac{u_{an,5}}{L+2L_{ac}} \\ -\frac{u_{ap,1}}{L} & \dots & -\frac{u_{ap,5}}{L} & -\frac{1}{L}u_{an,1} & \dots & -\frac{u_{an,5}}{L} \\ i_{ap}u_{ap,1} & \dots & i_{ap}u_{ap,5} & i_{an}u_{an,1} & \dots & i_{an}u_{an,5} \\ i_{ap}u_{ap,1} & \dots & i_{ap}u_{ap,5} & -i_{an}u_{an,1} & \dots & -i_{an}u_{an,5} \end{pmatrix}$

and $\boldsymbol{B}_{21} = \mathrm{diag}\left(\frac{i_{ap}}{C_{ap,1}}, \dots, \frac{i_{ap}}{C_{ap,5}}, \frac{i_{an}}{C_{an,1}}, \dots, \frac{i_{an}}{C_{an,5}}\right)$.

The features of the direct switched model inherits from the integrated current-energy model, which are multivariable, nonlinear and time-varying, and it extends with hybrid characteristics due to the switching nature of the involving switches. Furthermore, the two types of concerned dynamics, decomposed currents and numerous submodule voltages, are strongly coupled with each other. The instantaneous branch currents determine the voltage variations of the turn-on submodules, and meanwhile the selection of individual submodules to be switched on results in a unique influence to the branch current dynamics.

7.1.2 Analysis at steady state operating point

The analysis for the investigated single-phase MMC system at Steady State Operating Point (SSOP) provides an intuitive understanding of the current and voltage variables inside the MMC. Determination of SSOP is based on the separated continuous equivalent submodels of MMCC and the energy conservation law that links them together.

The active power delivered from the DC voltage sources is

$$p_{dc} = \frac{1}{2}u_{dc}i_{ap} + \frac{1}{2}u_{dc}i_{an} = u_{dc}i_{cir} \tag{7.10}$$

The active power is consumed by the resistance components in the MMC branches and in the AC-side load. The power consumption from the branch resistance is

$$p_{branch} = R(\frac{1}{2}i_{ph} + i_{cir})^2 + R(-\frac{1}{2}i_{ph} + i_{cir})^2 = \frac{1}{2}Ri_{ph}^2 + 2Ri_{cir}^2 \tag{7.11}$$

Consider the power at the AC-side RL load is $p_{AC} = \frac{1}{2}R_{ac}i_{ph}^2$, the following equation is obtained based on the energy conservation law

$$p_{dc} = p_{branch} + p_{AC} \tag{7.12}$$

$$u_{dc}i_{cir} = \frac{1}{2}Ri_{ph}^2 + 2Ri_{cir}^2 + \frac{1}{2}R_{ac}i_{ph}^2 \tag{7.13}$$

Suppose the phase current in SSOP is $i_{ph} = I_{ph}sin(\omega t + \phi)$. The circulating current at the SSOP is given as

$$i_{cir,ss} = \frac{u_{dc} - \sqrt{u_{dc}^2 - 2R(R + 2R_{ac})I_{ph}^2}}{4R} \tag{7.14}$$

The branch voltage in the steady state will be analyzed further. By applying Kirchhoff's voltage law (KVL) from one-phase MMC's AC side to the neutral point of the DC side, the positive and negative branch voltages can generally be expressed as

$$\begin{cases} u_{ap} &= \frac{1}{2}u_{dc} - L_{ac}\frac{d}{dt}i_{ph} - R_{ac}i_{ph} - Ri_{ap} - L\frac{d}{dt}i_{ap} \\ u_{an} &= \frac{1}{2}u_{dc} + L_{ac}\frac{d}{dt}i_{ph} + R_{ac}i_{ph} - Ri_{an} - L\frac{d}{dt}i_{an} \end{cases} \tag{7.15}$$

The branch voltage in the steady state can be obtained as

$$\begin{aligned} u_{ap,ss} &= \frac{1}{2}u_{dc} - (R_{ac} + \frac{R}{2})I_{ph}sin(\omega t + \phi) - \omega(L_{ac} + \frac{L}{2})I_{ph}cos(\omega t + \phi) \\ &= \frac{1}{2}u_{dc} - |Z_{eq}|I_{ph}sin(\omega t + \phi + \theta) \end{aligned} \tag{7.16}$$

where Z_{eq} is the equivalent impedance from the load and branch impedance satisfying that $Z_{eq} = (R_{ac} + \frac{R}{2}) + \omega(L_{ac} + \frac{1}{2}L)j$ and $|Z_{eq}| = \sqrt{(R_{ac} + \frac{R}{2})^2 + \omega^2(L_{ac} + \frac{1}{2}L)^2}$. θ is the equivalent impedance angle that $\theta = arctan\frac{\omega(L_{ac}+\frac{1}{2}L)}{R_{ac}+\frac{R}{2}}$. The branch power in steady state can therefore be obtained after certain arrangement as

$$p_{ap}(t) = \frac{1}{4}I_{ph}^2|Z_{eq}|cos(2\omega t + 2\phi + \theta) + \frac{1}{4}u_{dc}I_{ph}sin(\omega t + \phi) - I_{ph}|Z_{eq}|i_{cir,ss}sin(\omega t + \phi + \theta) \tag{7.17}$$

Then the branch energy $w_{ap}(t)$ is computed as the integration of the branch power and the averaging branch submodule voltage varying with time is

$$u_{SM,ss}(t) = \sqrt{\frac{2w_{ap}}{C_{SM}N_{SM}}} = \sqrt{\frac{2\int_0^T p_{ap}(t)dt}{C_{SM}N_{SM}}} \tag{7.18}$$

where T is the phase current period, N_{SM} is the number of branch submodules and C_{SM} is the submodule capacitance.

A SSOP analysis is given to a case study with these parameters: line frequency 50 Hz, branch resistance 0.01 Ω, branch inductance 2 mH, AC-side resistance 5 Ω, AC-side inductance 3.3 mH, DC-side voltage 250 V, submodule capacitance 3.3 mF, nominal submodule voltage 50 V and the AC-side current is controlled to 15 A. The calculated external currents, branch voltages, branch power and the averaging branch submodule voltage are shown in Fig. 7.3.

7.1.3 Discrete mathematical model

The proposed direct control method is formulated by evaluating the discrete mathematical developed in the last section. Define the sampling period T_{sw} and the discretization procedure can be implemented by applying the Euler forward equation:

$$\frac{d}{dt}x(t) \approx \frac{x(k+1) - x(k)}{T_{sw}} \tag{7.19}$$

where $x(k+1)$ and $x(k)$ are the value of the variable at time step $k+1$ and k, respectively. Therefore, the discrete dynamic model of the one-phase MMC can be obtained and will be used to predict the future current and energy states.

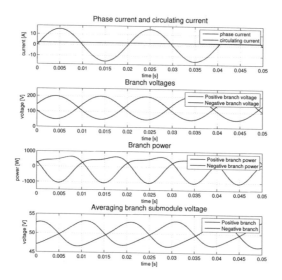

Figure 7.3: Circuit variables of one-phase MMC in SSOP

7.2 Problem formulation of direct control

7.2.1 Principle of model-based direct control

Compare with the conventional method that produces only the number of ON-switched submodules for each branch, direct control generates the optimal submodule switching states by referring to a switched dynamic model and minimizing a predefined cost function integrated with external current control, internal branch energy balancing and submodule voltage stability. The control diagram is shown in Fig. 7.4. An improved control performance of system variables, fast dynamic response and robustness for possibly varying submodule voltages can be expected. By extending the characteristic of the direct control method that controls each submodule separately, the variation of submodule capacitance and the arrangement of branch submodules with different operating voltages can also be handled by the detailed switched model.

The features of the proposed direct control can be summarized here.

- No intermediate stage for the branch voltage reference. Control decision for external current is obtained by taking the instantaneous submodule voltages into

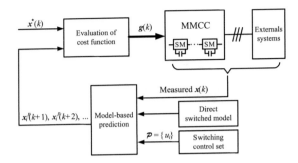

Figure 7.4: Basic control principle of MMCC direct control

consideration.

- Inherent treatment of parametric inhomogeneity of submodules. In the modeling layer of branch submodules the mathematical model of each submodule is individually developed with its own parameters, and the charging and discharging characteristics of the branch submodules can be exactly analyzed.

- Improved management of abnormal submodule voltages. Even if the submodule capacitor voltages are not yet balanced and oscillating, the external currents could still maintain in a good state.

- Extension to specific operation conditions. For example, the MMCC operation with branch submodules at different reference voltages can be realized. It is of great help when an MMCC is applied for an application with a low modulation index and a small number of branch submodules.

7.2.2 Formulation of subproblems in direct control

The three layers of a complete MMCC control framework are formulated into three subproblems, and the corresponding discrete-time model with discretization time T_{sw} and their cost functions are presented in the following paragraphs.

Control of current dynamics

The control of external and internal circulating currents will firstly be considered. Their discrete dynamic equations are

$$
\begin{cases}
i_{ph}(k+1) &= (-\frac{R+2R_{ac}}{L+2L_{ac}}T_{sw} + 1)i_{ph}(k) + \frac{T_{sw}}{L+2L_{ac}}(\sum_{i=1}^{5} u_{an,i}(k)g_{an,i}(k) \\
&\quad - \sum_{i=1}^{5} u_{ap,i}(k)g_{ap,i}(k)) \\
i_{cir}(k+1) &= (-\frac{R}{L}T_{sw} + 1)i_{cir}(k) - \frac{T_{sw}}{2L}(\sum_{i=1}^{5} u_{an,i}(k)g_{an,i}(k) \\
&\quad + \sum_{i=1}^{5} u_{ap,i}(k)g_{ap,i}(k)) + \frac{T_{sw}}{2L}u_{dc}
\end{cases}
\tag{7.20}
$$

Based on the measured currents $i_{ph}(k)$ and $i_{cir}(k)$ at the current time instant, their values at the next time instant can be predicted by (7.20) as $i_{ph}^{p}(k+1)$ and $i_{cir}^{p}(k+1)$. A cost function that penalizes the error between the prediction and the reference can be given as

$$
J_{current} = k_{ph}(i_{ph}^{p}(k+1) - i_{ph,ref}(k+1))^2 + k_{cir}(i_{cir}^{p}(k+1) - i_{cir,ref}(k+1))^2 \tag{7.21}
$$

where k_{ph} and k_{cir} are the two weighting factors for the current dynamic optimization. The reference value $i_{ph,ref}$ is usually given by users with the amplitude $I_{ph,ref}$ and frequency f_{ph} or calculated from the required AC-side power, and $i_{cir,ref}$ is computed based on SSOP analysis presented before.

Control of branch energy

In the layered control design procedure, the regulation of the sum and difference branch energy should be treated afterward. The discrete dynamic equations are given here:

$$
\begin{cases}
w_{\Sigma}(k+1) &= w_{\Sigma}(k) + \frac{i_{ph}(k)}{2}T_{sw}(u_{ap}(k) - u_{an}(k)) + i_{cir}T_{sw}(u_{ap}(k) + u_{an}(k)) \\
w_{\Delta}(k+1) &= w_{\Delta}(k) + \frac{i_{ph}(k)}{2}T_{sw}(u_{ap}(k) + u_{an}(k)) + i_{cir}T_{sw}(u_{ap}(k) - u_{an}(k))
\end{cases}
\tag{7.22}
$$

The cost function for the predicted branch energy variable w_{Σ}^{p} and $w_{\Delta^{p}}$ is

$$
J_{branch} = k_{\Sigma}(w_{\Sigma}^{p}(k) - w_{\Sigma,ref})^2 + k_{\Delta}(w_{\Delta}^{p}(k) - w_{\Delta,ref})^2 \tag{7.23}
$$

where k_{Σ} and k_{Δ} are the tuning weighting factors for the branch energy regulation. The reference value $w_{\Delta,ref}$ should be 0 and the $w_{\Sigma,ref}$ is calculated as the sum energy of all the submodules.

Balancing control of singular submodule voltage

The last issue is to consider the internal branch submodule balancing. The dynamic equations of submodule capacitor voltages are

$$
\begin{cases}
u_{ap,i}(k+1) &= u_{ap,i}(k) + \frac{T_{sw}}{C_{ap,i}}i_{ap}(k)g_{ap,i}(k) \\
u_{an,i}(k+1) &= u_{an,i}(k) + \frac{T_{sw}}{C_{an,i}}i_{an}(k)g_{an,i}(k)
\end{cases}
\tag{7.24}
$$

Then the cost function for the submodule voltages is formulated as

$$J_{sm} = k_{SM} \sum_{i=1}^{5} ((u_{ap,i}^{p}(k+1) - u_{SM,ref})^2 + (u_{an,i}^{p}(k+1) - u_{SM,ref})^2) \qquad (7.25)$$

7.2.3 Cost function integration

The final cost function is written as the sum of the ones in each control tasks as

$$J = J_{current} + J_{branch} + J_{sm} \qquad (7.26)$$

By introducing all the switching combination of the one-phase MMC to the cost function and find the optimal combination to minimize the cost function J, the regulated one-phase MMC can be regulated to its reference values. By tuning the weighting factors, the dynamic behavior of the corresponding state variables will be influenced. Finally, the one-step direct control problem can be formulated as

$$\begin{cases} \text{For the index number } i = 1, \, 2, \, ..., \, N_{sc} \\ \boldsymbol{x}_i^p(k+1) \;\; = \boldsymbol{A}_d \boldsymbol{x}(k) + \boldsymbol{B}_d(\boldsymbol{x}(k))\boldsymbol{u}_i(k) + \boldsymbol{v}(k) \\ J_i(k) \;\;\;\;\;\; = (\boldsymbol{x}_{ref}(k+1) - \boldsymbol{x}_i^p(k+1))^T \boldsymbol{K}(\boldsymbol{x}_{ref}(k+1) - \boldsymbol{x}_i^p(k+1)) \\ \boldsymbol{u}^*(k) \;\;\;\; = \underset{\boldsymbol{u}_i(k)}{\arg\min} \, J_i(\boldsymbol{x}(k), \boldsymbol{u}_i(k)) \end{cases} \qquad (7.27)$$

where i is the ordered switching combination of the one-phase MMC and N_{sc} is the total number of the possible switching combinations. For the chosen one-phase MMC, $N_{sc} = 2^{10} = 1024$, which means within each control period totally 1024 times of the cost function $J_i(k)$ should be performed. More generally, Problem (7.27) can be extended with a prediction horizon N and control input penalization:

$$\begin{cases} \text{For the index number } i = 1, \, 2, \, ..., \, N_{sc}^{N} \\ \boldsymbol{x}_i^p(k+1) = \boldsymbol{A}_d \boldsymbol{x}(k) + \boldsymbol{B}_d(\boldsymbol{x}(k))\boldsymbol{u}_i(k) + \boldsymbol{v}(k) \\ J_i^N(k) = \sum_{j=1}^{N}[(\boldsymbol{x}_{ref}(k+j) - \boldsymbol{x}_i^p(k+j))^T \boldsymbol{Q}(\boldsymbol{x}_{ref}(k+j) - \boldsymbol{x}_i^p(k+j)) \\ \;\;\;\;\;\;\;\;\; + \boldsymbol{u}_i^T(k+j-1)\boldsymbol{R}\boldsymbol{u}_i(k+j-1)] \\ \boldsymbol{u}^*(k) = \underset{\boldsymbol{u}_i(k),...,\boldsymbol{u}_i(k+N-1)}{\arg\min} \, J_i(\boldsymbol{x}(k+j), \boldsymbol{u}_i(k+j)) \end{cases} \qquad (7.28)$$

This generalized concept of implementing the prediction horizon of N, which is also N-step prediction model, can be depicted in Fig. 7.5. Each single-step optimization follows

$$\boldsymbol{u}_i(k) \in \mathcal{P} \subseteq \mathcal{U} \qquad (7.29)$$

where $\mathcal{P}$ is the possible control set at the current step and $\mathcal{U}$ is the total control set with size$(\mathcal{U}) = N_{sc}$. It is necessary to solve the optimization problem using multi-step policy in order to achieve an improved performance and system stability, particularly for the submodule voltages. However, compared with single-step method, multi-step optimization cannot be easily solved analytically due to the system nonlinearity and complexity. A search and decision strategy is required to find the minimal value of the cost function by examining among all the possible control candidates across the N prediction steps.

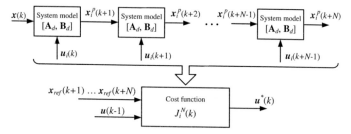

Figure 7.5: Scheme of direct control with the prediction horizon N

The following points will be concerned for the proposed search and decision strategy:

- Algorithm efficiency and feasibility. The computational complexity grows exponentially with the length of the prediction horizon and the manipulated states. The formulated optimization problem must be solved within each control period T_{sw}. In order to facilitate the demanding direct control method, efficient algorithms should be employed for solving the optimization problem and certain tradeoff needs to be made between the prediction horizon and computational cost.

- Optimal solution or sub-optimal one. It is understood that the computational time to obtain the optimal solution from direct control problem could be much longer than the usual sampling period of fast power electronics system. A quickly-generated suboptimal solution in most practical cases is more useful than the optimal one requiring unrealistic solution time.

- The resulting control performance for MMCC systems. Generally speaking, a longer prediction horizon in direct control method improves the system stability as well as control performance. Moreover, the receding horizon policy is repeated at the beginning of each control period, adding a high degree of robustness for the controlled system. It is expected that the MMCC system with direct controller gains an improved plant performance, robustness and flexible operating conditions.

7.2.4 Analysis of switching redundancy in MMCC

This subsection illustrates the high redundancy in MMCC direct control strategy by taking one-phase MMC with 5 submodules per branch as example. In order to analyze the discrete AC external voltage levels, the branch submodules are assumed to maintain the equal and constant voltage. Table 7.1 presents the relationship between numbers of ON-state submodules in two branches and the external DC and AC voltage levels. Since there exists six different voltage levels for each branch ranging from 0 to 5, and therefore, totally $6 \times 6 = 36$ different voltage levels can be observed at MMC DC and AC terminals. Actually, not all the 36 voltage levels are necessary for MMC operation. Under the assumption to avoid overly rapid submodule charging or discharging from the DC-side source, the feasible DC voltage levels are limited in the set $\{4, 5, 6\}$. In this case only 16 sets of voltage levels from 36 ones are utilized.

Table 7.1: Relation between number of ON-state branch submodules and external voltage levels

n_{AP}	5	5	5	5	5	5	4	4	4	4	4	4	3	3	3	3	3	3
n_{AN}	0	1	2	3	4	5	0	1	2	3	4	5	0	1	2	3	4	5
n_{DC}	5	6	7	8	9	10	4	5	6	7	8	9	3	4	5	6	7	8
n_{AC}	-2.5	-2	-1.5	-1	-0.5	0	-2	-1.5	-1	-0.5	0	0.5	-1.5	-1	-0.5	0	0.5	1
n_{AP}	2	2	2	2	2	2	1	1	1	1	1	1	0	0	0	0	0	0
n_{AN}	0	1	2	3	4	5	0	1	2	3	4	5	0	1	2	3	4	5
n_{DC}	2	3	4	5	6	7	1	2	3	4	5	6	0	1	2	3	4	5
n_{AC}	-1	-0.5	0	0.5	1	1.5	-0.5	0	0.5	1	1.5	2	0	0.5	1	1.5	2	2.5

The analysis of external voltage levels only considers the number of ON-state branch submodules and ignores which ones are switched on. One set of external voltages could respond to certain number of switching states for branch submodules, which are listed in Table 7.1. For example, for the case of $n_{DC} = 5$ and $n_{AC} = -1.5$ there are 25 different switching states for all the 10 submodules. The number of the switching states, e.g. switching redundancy, varies with external voltage levels. As the AC voltage level approaches close to its boundary values (± 2.5), switching redundancy decreases, meaning a reduced control set and potentially faster solution of direct control formulation (7.28). However, it still needs to deal with a huge number of redundancies when the AC voltage levels are around 0, which constitutes the main obstacle for real-time implementation and prediction horizon extension.

Table 7.2: Possible switching states (SSs) with respect to the AC external voltage levels

n_{AC}	-2.5	-2		-1.5	-1		-0.5	0		0.5	1		1.5	2		2.5
n_{DC}	5	4	6	5	4	6	5	4	6	5	4	6	5	4	6	5
Possible SSs	1	5	5	25	50	50	100	100	100	100	50	50	25	5	5	1
Total SSs	1	10		25	100		100	200		100	100		25	10		1

7.3 Summary

This chapter presents the methodology of developing direct switched model of MMCC by taking a one-phase MMC (MMCC-$\mathcal{K}_{2,1}$) as example. The direct switched model provides a thorough understanding of MMCC system from the external layer, the branch layer to the branch submodule layer. With the help of the direct switched model, a straightforward relationship is established between the submodule switching states and the MMCC current variables and the parameter variations of submodules can also be specifically handled. The second main content in this chapter is to formulate the model-based direct control problem in an optimization format, which uniformly controls the current, branch energy and submodule voltage under an integrated cost function. The greatest challenges of direct control for multivariable MMCC system are to enhance the solution feasibility for real-time implementation and to increase the prediction horizon for better control performance. In the end, the possible switching combinations are classified and the switching redundancy for each combination is analyzed. The solution schemes for direct control problem, which will be proposed in the next chapter, aim to deal with redundancy by making efforts from the both aspects: one is to improve algorithm in mathematics and another is to reduce the possible control set from a physical consideration.

8 Solution schemes of direct control method

Under the framework of model-based MMCC direct control, the control challenges inheriting from the continuous integrated model can effectively be handled by evaluating the discrete-time equations recursively based on the initial system states and a fixed number of possible control choices. However, the direct switched model incorporates among continuous-varying variables, e.g., the current and branch energy governed by differential or difference equations, and discrete-valued variables, e.g., submodule switching states, which is categorized as Hybrid System or Mixed Logical Dynamical (MLD) System. To handle the hybrid nature of MMCC, the commonly-used control structure requires a modulation block (modulator) to "translate" the continuous branch voltage commands into submodule switching signals, which introduces time-delay (related to the carrier frequency) as well as control variation (submodule voltage ripples are neglected). The direct control formulation offers a novel springboard for controlling an MMCC without presence of intermediate modulator, leading to discrete optimization that minimizes a real-valued cost function in a finite set of feasible solutions. Of course, discrete optimization problem can be straightforwardly solved by enumerating all feasible solutions, but it fails to meet the real-time requirement. The challenge of discrete optimization is to develop algorithms that are more efficient and practical so that it meets the requirement of controlling MMCC.

8.1 Tasks and specification of MMCC direct control

Control tasks of MMCC direct control
Generally speaking, the tasks of direct control method for MMCC systems can be classified into three subquests, which have different dynamics as well as required prediction horizons. (1) Reference tracking of current variables. The current dynamics in power electronics always requires a fast control response to fit varying operating conditions. It can already achieve a satisfactory performance without extending the prediction horizon greatly when the load time constant is small. (2) Branch submodule voltage regulation

and balancing. The key point is to regulate the DC averaging value of the oscillating submodule voltage to its nominal reference. To achieve the DC offset control, a long prediction, for instance, one period of line voltage, is preferred since the submodule voltage ripple has the same frequency as the line frequency. (3) Limitation of switching frequency and balancing control of switching frequency among the branch submodules. In the theoretical analysis, this point requires a even longer time period to return a more precise value of the switching frequency of one submodule. However, it does not mean that the prediction range must be accordingly long and the submodule switching history can be recorded to adjust the switching priority. As a summary, (1) and (2) are the primary quests that demand an according number of prediction horizon to ensure a swift control response and balancing of voltage pressure. (3) is a side quest, but it is of practical significance to improve conversion efficiency and switching balancing.

Specification for solving MMCC discrete optimization
For hybrid systems of the conventional power electronics devices, e.g. two- and three-level converter, direct control schemes are proposed based on solving a discrete optimization problem for a given prediction horizon and adopts the solution in a receding horizon policy. When direct control schemes are applied for MMCC systems, the theoretical challenges are to deal with the huge total control set and to obtain a feasible solution for real-time requirement. Two factors, e.g., the number of branch submodules N_{SM} and the prediction horizon N, influence greatly the dimension of the direct switched model, the size of the total control set and therefore the algorithm complexity. Thus, it is almost an impossible mission that applies direct control method to a high-voltage application requiring 100 or even 200 branch submodules or a excessively-high switching frequency case demanding a swift online computation. In this thesis direct control method focuses on a MMCC-based medium-voltage application, which have a limited number of branch submodules ($N_{SM} \leq 15$) determined by the medium-voltage scenario and a low control frequency ($f_{control} \leq 5\ kHz$) due to switching loss reduction. On the other hand, the complexity of the direct control problem grows exponentially with the length of the prediction horizon. A horizon of one is believed to suffice for current control and the use of longer horizons offers only a limited improvement [73]. However, as to the aspect of regulating the submodule voltage ripples, an appropriately long horizon can achieve a balanced voltage control while preventing unnecessary switching behaviors.

8.2 Implicit methods

The actuating variables in direct switched model is not continuous for taking any value, instead their values can only be chosen from several discrete number, e.g. $\{0, 1\}$ for

HB-SM and $\{-1, 0, 1\}$ for FB-SM. On this occasion, the conventional solution scheme of single-step optimization for continuous actuating variables is no longer feasible. Multi-step discrete optimization should be applied to solve the direct control problem. Nonlinearity introduced by the actuating variables leads that the multi-step optimization can no longer be handled analytically. First of all, search and decision strategy by examining all the control possibilities is introduced.

8.2.1 Exhaustive search

A straightforward solution method for the investigated multi-step optimization is to evaluate the values of the cost function under all the discrete control possibilities for the accounted N steps, which enumerates the complete possibility and can therefore always present the global minimal solution. Since this method examines all the possibilities in the total control set $\mathcal{U}$ regardless of computational cost and without any simplification, so it is termed as exhaustive search. However, for such multivariable systems as MMCCs with numerous switching combinations, the computational time is the main barrier for a practical implementation. For example, in the considered one-phase MMC, totally $2^{10} = 1024$ different switching possibilities exist for one-step prediction and the value increases to 1024^N when the prediction step rises. The computational effort increases exponentially with respect to the prediction horizon. Furthermore, each switching combination must be evaluated in 14 equations including current, energy and submodule voltage variables, adding the implementation difficulties. The exhaustive search is not a feasible method for solving the direct control problem, even sometimes for one-step control.

8.2.2 Dynamic programming

Dynamic programming is proposed to improve the algorithm efficiency of direct control. The key feature of dynamic programming is to divide the optimization problem into multiple stages and to determine the optimal decision from a finite number of decisions in a multistage way. The theoretical basis of dynamic programming is based on the principle of optimality. The principle of optimality can be described as "whatever the initial state and initial decision are, the remaining decisions must constitute an optimal policy with regard to the state resulting from the first decision [74]".

The discrete-time dynamic system of MMCC is given here.

$$\boldsymbol{x}(k+1) = \boldsymbol{f}_k(\boldsymbol{x}(k), \boldsymbol{u}(k)) \tag{8.1}$$

where the system equations $\boldsymbol{f}_k$ can be both nonlinear and time-varying, and $\boldsymbol{u}(k)$ corresponds to one switching combination of all the internal submodules. At time k, $\boldsymbol{u}(k)$ is chosen from a control constraint set $\mathcal{U}$ as $\boldsymbol{u}(k) \in \mathcal{U}$. For a fixed MMCC structure the control set $\mathcal{U}$ remains always the same. The cost function J accumulating over time from time $k = 0$ to $k = N$ can be given as

$$J_0(\boldsymbol{x}_0) = V_N(\boldsymbol{x}(N)) + \sum_{k=0}^{N-1} V_k(\boldsymbol{x}(k), \boldsymbol{u}(k)) \tag{8.2}$$

where $V_N(\boldsymbol{x}(N))$ is the terminal cost for the states at the end time N. In order to determine the optimal switching sequence $\boldsymbol{u}_{seq}^*(0)$ at time 0 for the above problem (8.2), which is

$$\boldsymbol{u}_{seq}^*(0) = \{\boldsymbol{u}^*(0),\ \boldsymbol{u}^*(1), ...,\ \boldsymbol{u}^*(N-1)\} \tag{8.3}$$

The principle of optimality will be introduced and the original discrete optimization problem from 0 to N is divided into subproblems. Assume that $\boldsymbol{u}_{seq}^*(0)$ is applied and a certain state $\boldsymbol{x}_i$ is achieved at time $k = i$. The remaining work is to solve this subproblem so that the cost-to-go from time i to N is minimized. The cost-to-go at state $\boldsymbol{x}(i)$ and time i is denoted as $J_i(\boldsymbol{x}_i)$. The truncated sequence $\boldsymbol{u}_{seq}(i) = \{\boldsymbol{u}^*(i),\ \boldsymbol{u}^*(i+1), ...,\ \boldsymbol{u}^*(N-1)\}$ is the optimal solution for the subproblem (8.3). The algorithm of dynamic programming can be written in the form of functional equation as

$$J_i^*(\boldsymbol{x}(i)) = \min_{\boldsymbol{u}(i)}[V_i(\boldsymbol{x}(i), \boldsymbol{u}(i))] + J_{i+1}^*(\boldsymbol{x}^*(i+1)) \tag{8.4}$$

(8.4) presents a solution policy that starts from the last stage J_{N+1}^*, extends to the tail subproblem J_N^* involving the last two stages, continues stage by stage and finally obtains the solution for the entire problem.

Dynamic programming can increase the computational speed, but it requires a huge memory to store the intermedia calculation results. Consider a quite simple case of half-bridge inverter that regulates the AC-side current i_{ph} with 4 possible switching combinations. The AC current starts from the same initial state $i_{ph}(0)$. Table 8.1 shows a comparison of the number of stepwise cost evaluation required for the exhaustive search and for the dynamic programming. It can be seen that dynamic programming can indeed reduce the calculation time for conventional converter application. However, a straightforward application of dynamic programming cannot satisfy real-time requirement of direct control. It must be combined with other solution schemes.

8.2.3 Other enumerative methods

Other implicit methods, such as speed-up strategy [75], branch and bound [76], cutting-plane method [77, 78] and expansion strategy [79], can be introduced to solve discrete

Table 8.1: Comparison of dynamic programming and exhaustive search

Number of the prediction horizon N	Number of cost evaluation by exhaustive search	Number of cost evaluation by DP	Calculation saving
1	4	4	0%
2	32	20	37.5%
3	192	84	56.3%
4	1024	340	66.8%
N	$N4^N$	$\frac{4}{3}(4^N - 1)$	$1 - \frac{4}{3N} + \frac{4}{3N4^N}$

optimization problems. It should be emphasized that these improved algorithms cannot well handle high-dimensional dynamic MMCC systems and provide real-time solution within one sampling interval for power electronics application.

8.3 Control set reduction

After analyzing the implicit methods of discrete optimization arising from MMCC direct control, it can be seen that these methods cannot well address the challenging issues of high system dimension and a huge total control set. It shows that the attempt simply from the mathematical aspect fails to produce a practical optimal solution within the sampling period. The next attempt focuses on developing a reduced control set (RCS) from the total control set by imposing physical switching constraints. In this section the principle of control set reduction is explained and fast RCS method is proposed.

8.3.1 Principle of control set reduction

Exhaustive search adopts the total control set with 1024 switching states and searches for the optimal one. Consider the background of medium-voltage application, restriction of submodule switching behavior is demanded in order to reduce energy loss and prolong semiconductor lifespan. The control set can actually be refined by introducing (1) physical consideration of avoiding unnecessary switching, (2) generation of current switching state referring to the last one and even (3) prediction for future switching states. Reduced control set method is proposed based on points (1) and (2). A hard constraint is given to the switching policy that at most one submodule switching state is allowed to be altered at the next time instant. Therefore, a reduced control set is organized based on the last switching state. Fig. 8.1 illustrates derivation of the reduced control set at time instant k. The control set is composed of three kinds of switching states:

1. $\delta n_{br}(k) = 1$. In one MMC branch, one OFF-state submodule is switched on.

2. $\delta n_{br}(k) = 0$. The last switching state $u(k-1)$ is always included in the current control set.

3. $\delta n_{br}(k) = -1$. One ON-state submodule is switched off.

Instead of searching the optimal switching state from 1024 ones, reduced control set method allows a control set $\mathcal{P}(k)$ with a fixed size, e.g., $N_{SM} + 1$. It can be extended in a quite straightforward way with the discrete dynamic programming method to reduce the times of cost function evaluation for a prediction horizon of more than one.

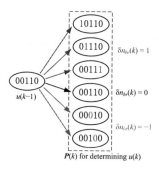

Figure 8.1: Derivation of current control set based on RCS

By limiting switching times, it brings an increased voltage difference among the branch submodules, causing a unevenly-distributed voltage pressure in the branch. Another drawback of RCS method is that a quick dynamic response to a sudden reference change is also limited by the switching principle. Regarding to the first issue, a certain number N_{bal} of submodules are forced to change their switching states at beginning of each control period [80]. The original control set with $N_{SM} + 1$ elements has to expand to $1 + C^{1}_{N_{SM}} C^{N_{bal}}_{N_{SM}-1}$. To cope with the second drawback, e.g., to achieve a swift dynamic response, the variation index $\delta n_{br}(k)$ will be assigned with a larger integer rather than 0, 1 or -1 according to control requirements, so that future switching states must be predicted. Fast RCS method is based on RCS method and combined with prediction $\delta n_{br}(k)$.

8.3.2 Fast RCS method

Fast RCS method follows the same principle that switches branch submodules only when necessary, but it combines a prediction for variation of ON-state branch submodules and in most cases leads to a smaller control set. Its control configuration is shown in Fig. 8.2.

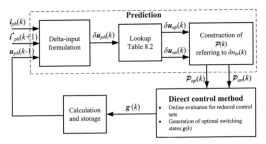

Figure 8.2: Functional blocks of fast RCS

Control set establishment based on prediction

A prediction for $\delta n_{br}(k)$ can be used to further compress the control set from RCS method. Fig. 8.3 shows two examples. Fig. 8.3(a) describes a similar situation as Fig. 8.1, but it is known in advance that $\delta n_{br}(k) = 1$ has to be applied for the current time instant. It results in a control set with only 3 elements. If $\delta n_{br}(k) = 2$ and the last state is "00110", the control set $\mathcal{P}(k)$ has only 3 elements.

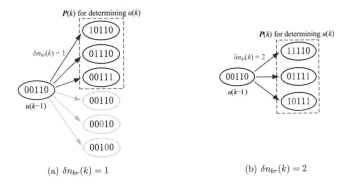

(a) $\delta n_{br}(k) = 1$ (b) $\delta n_{br}(k) = 2$

Figure 8.3: Control set derivation based on fast RCS

Prediction based on delta-input formulation

The dynamic equation of external current can be expressed in a delta-input form with

$\delta u_{ph}(k)$ as the new control input:

$$
\begin{aligned}
i_{ph}(k+1) &= ai_{ph}(k) + bu_{ph}(k) \\
&= ai_{ph}(k) + b(u_{ph}(k-1) + \delta u_{ph}(k)) \\
&= ai_{ph}(k) + bu_{ph}(k-1) + b\delta u_{ph}(k)
\end{aligned}
\tag{8.5}
$$

where a and b are the involved electrical parameters. It can be solved that

$$
\delta u_{ph}(k) = \frac{1}{b}[i_{ph}(k+1) - ai_{ph}(k) - bu_{ph}(k-1)]
\tag{8.6}
$$

To reach the reference $i_{ph}^*(k+1)$, (8.6) is rewritten as

$$
\delta u_{ph}(k) = \frac{1}{b}[i_{ph}^*(k+1) - ai_{ph}(k) - bu_{ph}(k-1)]
\tag{8.7}
$$

The discrete level $\delta n_{ph}(k)$ of delta-phase-voltage is obtained after dividing $\delta u_{ph}(k)$ by normal submodule voltage $U_{SM,nom}$ and rounding operation.

$$
\delta n_{ph}(k) = \frac{1}{2}\mathrm{round}(\frac{2}{bU_{SM,nom}}[i_{ph}^*(k+1) - ai_{ph}(k) - bu_{ph}(k-1)])
\tag{8.8}
$$

Based on delta-phase-level $\delta n_{ph}(k)$ the integer values of delta-branch-number $\delta n_{ap}(k)$ and $\delta n_{an}(k)$ are calculated as

$$
\delta n_{ph} = \frac{1}{2}(\delta n_{an} - \delta n_{ap})
\tag{8.9}
$$

So far, the variation number for branch submodules is determined to realize external current tracking. Based on the principle that minimizes the switching behavior of branch submodules, Table 8.2 is obtained for determining the delta-branch-number δn_{ap} and δn_{an}. By looking up the corresponding delta-input δn_{br}, it results in the following three cases:

- if $\delta n_{br} > 0$ and $|\delta n_{br}|$ larger than the number of OFF-state submodules, then switch all of them on. The size of the control set is 1;

- if $\delta n_{br} < 0$ and $|\delta u_{br}|$ larger than the number of ON-state submodules, then switch all of them off. The size of the control set is 1;

- else δn_{br} is in the range of OFF- or ON-state submodules. The number of the elements in the reduced control set is given in Table. 8.3.

By examining the numbers listed in Table 8.3, it can be seen that in most cases fast RCS method has control sets less than 6. The reason is that fast RCS method refers to not only the history information $u(k-1)$ from the last step but also the predicted information $\delta n_{br}(k)$.

Table 8.2: Determination of delta-branch-number

δn_{ph}	$-U_{max}$	...	$-\frac{3}{2}$	-1	$-\frac{1}{2}$	0	$\frac{1}{2}$	1	$\frac{3}{2}$	...	U_{max}
δn_{ap}	$\text{ceil}(U_{max})$	...	2	1	1	0	0	-1	-1	...	$-\text{floor}(U_{max})$
δn_{an}	$-2U_{max} + \text{ceil}(U_{max})$	...	-1	-1	0	0	1	1	2	...	$2U_{max} - \text{floor}(U_{max})$

Table 8.3: Control set size with respect to delta-branch-number

	Number of ON-state submodules $u(k-1)$	0	1	2	3	4	5
Number of elements in the RCS when	$\delta n_{br}(k) = -5$	0	0	0	0	0	1
	$\delta n_{br}(k) = -4$	0	0	0	0	1	5
	$\delta n_{br}(k) = -3$	0	0	0	1	4	10
	$\delta n_{br}(k) = -2$	0	0	1	3	6	10
	$\delta n_{br}(k) = -1$	0	1	2	3	4	5
	$\delta n_{br}(k) = 0$	1	1	1	1	1	1
	$\delta n_{br}(k) = 1$	5	4	3	2	1	0
	$\delta n_{br}(k) = 2$	10	6	3	1	0	0
	$\delta n_{br}(k) = 3$	10	4	1	0	0	0
	$\delta n_{br}(k) = 4$	5	1	0	0	0	0
	$\delta n_{br}(k) = 5$	1	0	0	0	0	0

8.3.3 RCS methods with extended prediction horizon

The size of control set and the evaluation times of cost function are major criteria indicating practicability. Table 8.4 shows the sizes of the control sets for RCS method and fast RCS method regarding to direct control of one-phase MMC. As a benchmark the case of exhaustive search is also included.

Since exhaustive search examines the total control set with 1024 switching combinations for each prediction step, the obtained solution is naturally the global minimum and suits for various operating conditions. However, the computational amount increases exponentially with a base value 1024, and it becomes already an impossible calculation task when the prediction step is set to 2. The control set size from RCS method is always 6 for one branch and 36 for one phase. It is still a feasible and effective choice to realize a direct control formulation with a prediction horizon less than 3. Fast RCS method actually offers the best feasibility due to prediction of future switching sequence. Fast RCS method is promising for online implementation with a longer prediction horizon. Although RCS methods can easily be integrated with discrete dynamic programming to further improve the algorithm efficiency, it is still impossible to extend the prediction horizon to a sufficiently long range due to larger computation amount and higher storage expense. RCS and fast RCS cannot provide global minimal solution, but simulation and hardware verification shows that their solutions are stable for one-phase MMC system.

Table 8.4: Sizes of control sets and evaluation times for RCS-class methods

	Control set size		Evaluation times when horizon 1		Evaluation times when horizon 5		Optimal solution?
	Best	Worst	Best	Worst	Best	Worst	
Exhaustive search	1024	1024	1024	1024	10^{15}	10^{15}	Global minimum
RCS	6	6	6^2	6^2	10^7	10^7	Not global minimum
fast RCS	1	10	1	25	10^2	10^3	Not global minimum

8.4 Event-based direct control method

Since the emergence of embedded microprocessors and rapid development computer technology, the discrete-time systems are becoming increasingly important in many diverse applications. In the common time-triggered control scheme, the controller could adjust the submodule switching states at every sampling instant, which would inevitably cause frequent changes of the submodule switching state and therefore leads to unnecessary switching losses as well as high computational demand. For this reason, a event-based approach is proposed to overcome these disadvantages involving in MMCC control. In the event-based strategy, the submodules are switched only when certain conditions are satisfied in the event-based formulation. The features of event-based direct control scheme are stabilized MMCC operation, increased switching efficiency and unified maximal voltage balancing. As a first step, the energy variation in one branch is analyzed.

8.4.1 Voltage variation analysis of branch submodules

The case of one-phase MMC with 5 submodules per branch is used here as an illustrative example to analyze interaction among branch voltage, branch current and submodule voltage variation. A numerical simulation is implemented for the investigated MMC by employing a direct control for the external current. The external current is controlled to a reference sinusoid (50 Hz and 20 A) with a prediction horizon 3 and the control period is set to 1 ms, which contains 20 control decisions in one period. Fig. 8.4 shows the simulation results, where the first subplot records branch voltage waveform, e.g. the number of ON-state submodules $n_{ap(ON)}$ in the positive branch AP. The identical blue curves in the subsequent five subplots are the same branch current i_{ap}. By examining i_{ap} and $n_{ap(ON)}$, the amounts of charged and discharged energy in each time duration can be visualized, which are represented by the area shaded in red (submodule charged region) and blue (submodule discharged region), respectively. For instance, between the time instant 11 and 12, where the branch current approaches the maximum negative value and four submodules are required to be turned on, then these four submodules will be discharged simultaneously due to the common discharging current. The sum of discharged energy can equivalently be evaluated by adding up the five blue-shaded areas from subplots $n_{ap(ON)} = 1$ to $n_{ap(ON)} = 5$. Similarly, the sum of charged energy can be calculated by adding up the three red-shaded areas from subplots $n_{ap(ON)} = 3$ to $n_{ap(ON)} = 5$. Therefore, the total amount of energy variation in Branch AP can explicitly be computed within each time step. The profile of energy variation as well as the voltage ripple can be computed.

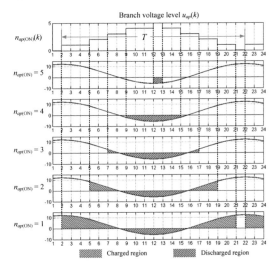

Figure 8.4: Illustration of energy charge and discharge for branch submodules (starting from $u_{ap(ON)}(1) = 0$)

Another interesting fact is that the branch current has a longer time duration and a higher current amplitude in the charging region than in the discharging region, which could result in persistive submodule voltage rising. The reason why submodule voltages stay stable is that less number&times submodules are enforced to turn on in the charged region compared with in the discharged region. As shown in Fig. 8.4, totally 19 number&times ON-states of branch submodules are observed in the charging cycle while up to 34 number&times exist in the discharging cycle, so that the total increased energy ΔQ_{charge} and the total decreased energy $\Delta Q_{discharge}$ for branch submodules remain the same. Meanwhile, it is also more challenging to achieve submodule voltage regulation or balancing in the charging region because less number&times of switches are permitted than in the discharging region.

8.4.2 Submodule voltage prediction

Based on the analysis of branch energy variation, this subsection proposes a model-based prediction method for submodule voltage variation, which can efficiently be applied for a online prediction.

Refer to (7.1) and (7.14), the branch current $i_{ap,ss}$ in steady state is expressed as

$$i_{ap,ss}(t) = \frac{1}{2}I_{ac}\sin(\omega t + \phi) + i_{cir,ss} \tag{8.10}$$

Assume that SMi in Branch AP is switched on at time instant kT_{sw} and remains inserted for a time span $k_{on}T_{sw}$, e.g. $g_{ap,i}(t) = 1$ for $t \in [kT_{sw} \quad (k+k_{on})T_{sw}]$, the voltage variation $\Delta u_{ap,i}(k+k_{on})$ after this time span can be computed based on (7.8) as

$$
\begin{aligned}
\Delta u_{ap,i}(k+k_{on}) &= \int_{kT_{sw}}^{(k+k_{on})T_{sw}} \frac{1}{C_{ap,i}} g_{ap,i}(t) i_{ap}(t) dt \\
&= \frac{1}{C_{ap,i}} \int_{kT_{sw}}^{(k+k_{on})T_{sw}} (\frac{1}{2}I_{ac}\sin(\omega t + \phi) + i_{cir,ss}) dt \\
&= -\frac{1}{2\omega C_{ap,i}} I_{ac}\cos(\omega t + \phi) \Big|_{kT_{sw}}^{(k+k_{on})T_{sw}} + \frac{1}{C_{ap,i}} i_{cir,ss} k_{on} T_{sw}
\end{aligned} \tag{8.11}
$$

The first term in (8.11) can be transformed by the Sum-to-Product Identity into a compact form as

$$
\begin{aligned}
\Delta u_{ap,i}(k+k_{on}) &= \frac{I_{ac}}{\omega C_{ap,i}} \sin(\frac{\omega T_{sw}}{2}k_{on}) \sin(\omega T_{sw}(k + \frac{1}{2}k_{on}) + \phi) + \frac{T_{sw}}{C_{ap,i}} i_{cir,ss} k_{on} \\
&= \frac{1}{C_{ap,i}} [\frac{I_{ac}}{\omega} \sin(\frac{\omega T_{sw}}{2}k_{on}) \sin(\omega T_{sw}(k + \frac{1}{2}k_{on}) + \phi) + T_{sw} i_{cir,ss} k_{on}]
\end{aligned} \tag{8.12}
$$

Note that the diversity of submodule voltages from the same branch is dependent on the submodule capacitance in this prediction algorithm. After excluding the submodule capacitance $C_{ap,i}$ from (8.12), the left terms inside the square bracket are the same for all the branch submodules. It can be precalculated and stored in a look-up table with respect to the current instant k and the predicted ON-duration k_{on}.

For a predetermined SSOP, the submodule voltages can be predicted based on the turn-on instant kT_{sw} and the turn-on duration $k_{on}T_{sw}$, symbolized as $u_{ap,i}(k|k_{on})$ and $\Delta u_{an,i}(k + k_{on})$ for the both branches. The submodule voltage prediction is given as

$$\begin{cases} u_{ap,i}(k|k_{on}) &= u_{ap,i}(k) + \Delta u_{ap,i}(k + k_{on}) \\ u_{an,i}(k|k_{on}) &= u_{an,i}(k) + \Delta u_{an,i}(k + k_{on}) \end{cases} \tag{8.13}$$

In event-based method, the above equations are programmed for each submodule with respect to their individual parameters and applied to predict the number of maximal possible turn-on periods without exceeding the predefined safe range. The iterations for the term $\Delta u_{ap,i}(k + k_{on})$ with k_{on} increasing one by one can effectively be calculated by establishing a common look-up table and locating the exact value.

8.4.3 Control configuration

The configuration of event-based direct control method for MMCC is shown in Fig. 8.5. Each submodule is monitored by one event generator that analyzes the electrical information (capacitor voltage value $u_{aj,i}(k)$, the switching history $g_{aj,i}(k - 1)$ and the branch current $i_{aj}(k)$, $j = p$, n; $i = 1$, ..., 5.) and then generates a switching index to indicate its switching priority. This switching index takes the following two aspects into consideration:

1. its potential switching capability for a possibly longer ON-state duration without breaking the predefined voltage boundaries;

2. its probability of reaching the predefined boundaries as close as possible;

3. Other aspects regarding to switching frequency can also be modelled and included for further adaption.

After that a functional block collects the switching indexes from all the event generators and constructs a feasible control set for the present control period. The switching combinations in the set will be synthesized in the direct controller so that the optimal one can be determined. In this configuration the complex computation of realizing a long prediction horizon for submodule voltage regulation is removed from the centralized direct

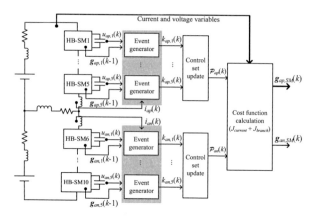

Figure 8.5: Event-based direct control method with extended prediction horizon

controller and realized in the event generators. These local event generators, referring to the according submodule models and the estimated voltage variation, will manage the submodule voltage within the safety boundaries and avoid unnecessary switching action by means of generation of switching index. In the following paragraphs the functionality of the involved blocks in event-based direct control method will be discussed in detail.

Estimation of submodule voltage ripple

According to Fig. 8.4 or the SSOP analysis based on the continuous equivalent model, the total branch energy increasing in the charging cycle and decreasing in the discharging cycle can be computed. Then the averaging submodule voltage variation can be calculated by (7.18) for any given AC current reference. Symbol the obtained upper bound and lower bound for the profile of the averaging submodule voltage as $U_{SM,upr}$ and $U_{SM,lowr}$, which will be applied as soft constraints for the submodule voltage regulation.

$$\begin{cases} U_{SM,upr} &= \sup(u_{SM,ss}(t)) \\ U_{SM,lowr} &= \inf(u_{SM,ss}(t)) \end{cases} \text{ for } t \in [0,\ T] \tag{8.14}$$

The estimated branch voltage profile in one period based on (7.18) is precise for each submodule only if the sorting algorithm for the branch submodules runs in a sufficiently high frequency. With the help of the online sorting and the frequently submodule switching, the submodule voltages can therefore rise and drop in a unified pace and achieve the same extreme values, as given in (7.18). However, this estimation is still valid to take as a reference for regulation of the submodule voltage variation.

Event generator

Event generator is the key structure in event-based direct control method, which evaluates the overall state of one submodule and determines its priority of being switched. The mechanism of event generator can be described as follow. The branch current is synchronized by detecting the zero-crossing of branch current and locking the phase angle of the phase current. Then the prediction equation (8.13) can be applied at any time instant k to predict the submodule voltage variation after any size of ON-state horizon k_{on}. However, in order to guarantee that the extreme submodule voltage is under regulation and avoid higher submodule voltage ripple, the evaluation of (8.13) is divided into two cycles, one is the positive branch current that charges ON-switched submodules and another is the negative discharging branch current.

- In the charging region the submodule voltage $u_{ap,i}(k)$ is predicted to obtain the maximal number of continuous ON-state intervals $k_{ap,i(on)}$ by setting the upper bound $U_{SM,upr}$ as constraint, which satisfies

$$
\begin{cases}
u_{ap,i}(k|k_{ap,i(on)}) & < \quad U_{SM,upr} \\
u_{ap,i}(k|k_{ap,i(on)} + 1) & > \quad U_{SM,upr}
\end{cases}
\tag{8.15}
$$

The determined index $k_{ap,i(on)}$ by (8.15) reflects the submodule capacity of being switched on without breaking the upper bound, termed as switching capacity index. In order to reduce the switching frequency and improve the conversion efficiency in the medium-voltage application, the submodule is once turned on and preferred to remain on for a possibly longer time.

Additionally, another index $k_{ap,i(U)}$ is introduced to evaluate the proximity between the upper bound, which is defined as

$$
k_{ap,i(U)} = \frac{U_{SM,upr} - u_{ap,i}(k|k_{ap,i(on)})}{U_{SM,ref}} > 0
\tag{8.16}
$$

The proximity index $k_{ap,i(U)}$ assists to make the decision when more than one submodules have the same possible ON-state duration, so that the submodule with the predicted end voltage further away from the upper bound will be selected with priority. In summary, in the charging region, the final switching index $k_{ap,i}$ can be defined as the summation of these two indexes.

$$
k_{ap,i} = k_{ap,i(on)} + k_{ap,i(U)}
\tag{8.17}
$$

A special case for $k_{ap,i}$ is that any further attempt of switching on the SMi will directly exceed the upper bound, leading to $k_{ap,i(on)} = 0$. The proximity index

$k_{ap,i(U)}$ in (8.16) has to be reformulated by computing the error between one-step prediction $u_{ap,i}(k|1)$ and $U_{SM,upr}$ as

$$k_{ap,i(U)} = \frac{U_{SM,upr} - u_{ap,i}(k|1)}{U_{SM,ref}} < 0, \text{ if } k_{ap,i(on)} = 0 \qquad (8.18)$$

so that the degree of exceeding the upper limit can be quantified. The algorithm in each event generator in the charging cycle is shown in Fig. 8.6. For a clear illustration, calculation of the switching index for Branch AP is summarized as:

$$k_{ap,i} = \begin{cases} \frac{U_{SM,upr} - u_{ap,i}(k|1)}{U_{SM,ref}} < 0 & \text{if } k_{ap,i(on)} = 0 \\ k_{ap,i(on)} + \frac{U_{SM,upr} - u_{ap,i}(k|k_{ap,i(on)})}{U_{SM,ref}} > 0 & \text{if } k_{ap,i(on)} \geq 1 \end{cases} \qquad (8.19)$$

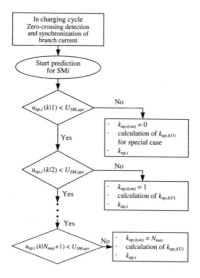

Figure 8.6: Flowchart of the event generator in charging cycle

- Similarly, in the discharging region, the corresponding indexes can be defined in the same way as (8.17). However, the involved factors $k_{ap,i(on)}$ and $k_{ap,i(U)}$ are defined relating to the lower bound $U_{SM,lowr}$. Their conditions and expressions are directly given here.

$$\begin{cases} u_{ap,i}(k|k_{ap,i(on)}) > U_{SM,lowr} \\ u_{ap,i}(k|k_{ap,i(on)} + 1) < U_{SM,lowr} \end{cases} \qquad (8.20)$$

and

$$k_{ap,i} = \begin{cases} \dfrac{u_{ap,i}(k|1) - U_{SM,lowr}}{U_{SM,ref}} < 0 & \text{if } k_{ap,i(on)} = 0 \\[3mm] k_{ap,i(on)} + \dfrac{u_{ap,i}(k|k_{ap,i(on)}) - U_{SM,lowr}}{U_{SM,ref}} > 0 & \text{if } k_{ap,i(on)} \geq 1 \end{cases} \qquad (8.21)$$

The switching indexes $k_{an,i}$ for the submodules in the negative branch AN can be obtained in the same manner. These indexes will be analyzed to formulate the feasible control set for the control decision at time k.

- An illustrative example for calculating the switching indexes for five positive branch submodules during the charging cycle is given in Fig. 8.7. In order to clearly show the calculation of the switching indexes, the five submodules have different initial capacitor voltages at the time instant k and their capacitor charging dynamics differ with each other due to MOSFET internal resistance and submodule capacitance. Since the prediction of submodule voltage is synchronized with the given AC current reference, the time instant $k + 4$ with the changed sign of i_{ap} can be identified, and therefore, the prediction horizon applied inside each event generator terminates at time $k + 4$. The initial submodule voltages and the predicted submodule voltage changes in the following time steps are plotted with solid dots and dashed lines in diverse colors.

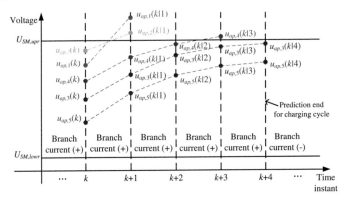

Figure 8.7: Example of determination of switching indexes

The corresponding switching indexes are calculated and listed in Table 8.5, and their relative sizes are given as $k_{ap,1}(k) < k_{ap,2}(k) < 0 < k_{ap,4}(k) < k_{ap,3}(k) < k_{ap,5}(k)$. The sign of the switching indexes also indicates the switching priority. If a certain number of the submodules is required to be switched on, the submodules with higher switching

Table 8.5: Values of the switching indexes at time k

	$k_{ap,i(on)}$	$k_{ap,i(U)}$	$k_{ap,i}$
SM-AP1	0	$\frac{U_{SM,upr}-u_{ap,1}(k|1)}{U_{SM,ref}}$	$\frac{U_{SM,upr}-u_{ap,1}(k|1)}{U_{SM,ref}}$
SM-AP2	0	$\frac{U_{SM,upr}-u_{ap,2}(k|1)}{U_{SM,ref}}$	$\frac{U_{SM,upr}-u_{ap,2}(k|1)}{U_{SM,ref}}$
SM-AP3	4	$\frac{U_{SM,upr}-u_{ap,3}(k|4)}{U_{SM,ref}}$	$4+\frac{U_{SM,upr}-u_{ap,3}(k|4)}{U_{SM,ref}}$
SM-AP4	2	$\frac{U_{SM,upr}-u_{ap,4}(k|2)}{U_{SM,ref}}$	$2+\frac{U_{SM,upr}-u_{ap,4}(k|2)}{U_{SM,ref}}$
SM-AP5	4	$\frac{U_{SM,upr}-u_{ap,5}(k|4)}{U_{SM,ref}}$	$4+\frac{U_{SM,upr}-u_{ap,5}(k|4)}{U_{SM,ref}}$

indexes are ought to be switched on, and meanwhile, if submodules should be switched off, the ones with lowest switching indexes are considered with priority. Compare with the classic voltage-sorting method only regarding to the present voltage values, this event-based method additionally considers the switching potential for the future steps, and is therefore capable of offering an optimized switching sequence.

Feasible control set update

There are two policies of updating control set.

a) Static case for updating control set
For the static case there exists three possible control choices, Choice I is to keep the value of the last branch voltage and swap the switching states of two submodules if necessary, Choice II is to turn on one submodule from the OFF-state ones and Choice III is to turn off one submodule.

1. Choice I ($\delta u_{br} = 0$) can directly be obtained from the switching states $u(k-1)$ in the last control period if all the switching indexes of ON-state submodules are above one. Else inter-switch two submodules for a better voltage regulation;

2. Choice II ($\delta u_{br} = 1$) is based on $u(k-1)$ and one additional submodule with the highest switching index among all the Off submodules;

3. Choice III ($\delta u_{br} = -1$) is also based on $u(k-1)$, but one of the ON-state submodules with the lowest switching index is switched off.

Thanks to the event generators, the feasible control set at this step is constrained to the size of 3 for one branch, which is already half of the set size from RCS method.

b) Prediction for updating control set
The prediction algorithm already proposed in Subsection 8.3.2 can also be employed to

enhance dynamic performance and compress the feasible control set.

Evaluation of cost function

Only a quite limited number of control choices from the feasible control set are required to be evaluated by the cost function. Moreover, the cost function is reduced to concern only the current and branch energy dynamics, since the computation for submodule voltage regulation is now distributed to local event generators. In the analyzed one-phase MMC case there exist only $3 \times 3 = 9$ possible switching combinations that are required to be computed in the order-reduced cost function. A adaptively long prediction horizon for submodule voltage regulation and balance is now ensured by the local event generators. Therefore, event-based direct control scheme offers the least computational requirement and achieves the balancing of extreme submodule voltages benefited from a long prediction horizon. Finally, the flowchart of event-based direct control algorithm is given in Fig. 8.8.

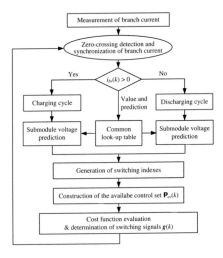

Figure 8.8: Flowchart of event-based direct control method

8.4.4 Comparison to voltage-sorting method

The conventional and effective switching principle is based on a voltage-sorting algorithm, which gives a priority for the submodules with lowest voltages to insert into the circuit when the branch current charges the switched-on submodules and the ones with highest voltages to insert when the branch current discharges the ON-SMs. Fig. 8.9(a) shows the resulting submodule voltages and their switching states. It can be seen that each submodule is switched on and off quite frequently to guarantee a uniformed voltage rise and drop. The voltage-sorting method can be regarded as a simple non-model-based one-step prediction, which requires merely the relative values of instantaneous submodule voltages and the branch current direction. It ensures a balanced submodule voltage distribution at each time instant at the expense of greatly increasing the switching frequency and frequently-demanded online computation.

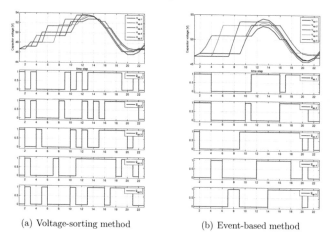

(a) Voltage-sorting method   (b) Event-based method

Figure 8.9: Comparison of submodule switching performance

By contrast, event-based method takes the branch current at the current instant and the future variation into account, and therefore it can avoid unnecessary switching as long as the submodule voltage lies within the precalibrated range. The resulting pulses of the event-based method is shown in Fig. 8.9(b). It can be seen that this method brings about a reduced switching frequency for the branch submodules while a well-regulated bound for their voltages. The submodule is switched on so long as its voltage in the following time instants would not exceed the boundary. Additional criteria can be included into this event algorithm, such as switching frequency and submodule aging rate.

8.4.5 Extension of direct control method

Case I: Switching compensation for parametric inhomogeneity

Since the capacitors in branch submodules cannot be identical in real practice, the charging and discharging profiles of submodule capacitors differ from each other, and therefore, different voltage variations will be observed for the ON-state submodules even though they are switched on in the same control period. Moreover, the other circuit parameters, such as the voltage drop across the power switch, the turn-on and turn-off delay, internal distributed resistance and inductance as well as the submodule initial voltage will all cause the switching dynamics different from the others. Therefore, in certain occasions the switching burden is demanded to be shared equally among branch submodules. In the framework of direct control method, this requirement can be mathematically modelled and integrated into the cost function or the event generator.

The switching times $n_{\text{SM}i}$ of ith submodule during the time range $N_{sw}T_{sw}$ is calculated as $n_{\text{SM}i}(k-1) = \sum_{j=1}^{N_{sw}} g_{\text{SM}i}(k-j)$, where $g_{\text{SM}i}(k-j) = 1$ or 0 indicating the submodule switching state and the last N_{sw} values are stored in the register. A cost term can be added to (7.26)

$$ J_{sw} = k_{sw} \sum_{i=1}^{N_{SM}} k_{nom} n_{\text{SM}i}(k-1) \tag{8.22} $$

where k_{sw} is the weighting factor for switching times balancing and k_{nom} is the normalizing factor for switching counts. Similar formulation can also be given to the event generator for generating the switching index considering switching times balancing.

Case II: Submodules with different operating voltages

An interesting application scenario of MMCC is that its branch is composed of two types of submodules, one with a higher DC offset voltage than another. This application scenario is quite helpful for medium-voltage DC transmission, which contains less number of branch submodule and requires a quite low modulation ratio. An increased number of low-voltage submodules can participate to generate a smooth AC voltage while the fixed number of high-voltage submodules sustain the medium DC voltage. For example, a medium-voltage application of three-phase MMC (MMCC-$\mathcal{K}_{3,2}$) realizing AC 550 V to DC 8 kV without employing a step-up transformer. Suppose each converter branch is composed of 12 submodules with rating voltage 400 V, and the AC-side voltage amplitude is around 700 V. Only ± 2 voltage levels can be realized at the AC side. If the branch is composed of two types of submodules (seven 200V-Submodules and five 800V-Submodules), the DC voltage 8 kV is satisfied by adopting the 800V submodules and the AC voltage can be synthesized by ± 4 voltage levels offered by 200V-Submodules.

In general, suppose there are two types of submodules, denoted as High-SM with high submodule voltage and Low-SM with low submodule voltage. Note that additional energy variables should be accordingly defined to evaluate the energy situation for both High-SMs and Low-SMs. Firstly, the branch energy is the energy sum of the High-SM and Low-SM, which are $w_{ap} = w_{ap(high)} + w_{ap(low)}$ and $w_{an} = w_{an(high)} + w_{an(low)}$

Then four new energy variables can be defined to evaluate their balance states as

$$
\begin{cases}
w_{\Sigma(high)} &= w_{ap(high)} + w_{an(high)} \\
w_{\Delta(high)} &= w_{ap(high)} - w_{an(high)} \\
w_{\Sigma(low)} &= w_{ap(low)} + w_{an(low)} \\
w_{\Delta(low)} &= w_{ap(low)} - w_{an(low)}
\end{cases}
\tag{8.23}
$$

Dynamic equations for the above four variables can be obtained similar to (7.5) as

$$
\begin{cases}
\frac{d}{dt}w_{\Sigma(high)} &= \frac{1}{2}i_{ph}\big(u_{ap(high)} - u_{an(high)}\big) + i_{cir}\big(u_{ap(high)} + u_{an(high)}\big) \\
\frac{d}{dt}w_{\Delta(high)} &= \frac{1}{2}i_{ph}\big(u_{ap(high)} + u_{an(high)}\big) + i_{cir}\big(u_{ap(high)} - u_{an(high)}\big)
\end{cases}
\tag{8.24}
$$

where $u_{ap(high)}$ is the branch voltage component from the High-SMs in Branch AP. By controlling the insert and bypass of the High-SMs the sum energy $w_{\Sigma(high)}$ and difference energy $w_{\Delta(high)}$ of the High-SMs are under regulation. The same dynamic equations can also be developed for Low-SMs. The system order of MMC with different operating voltages is increased in order to keep their respective reference voltages. However, the principle of designing direct control is the same.

8.5 Summary

This chapter discusses the solution schemes of direct control method for direct switched model. It is known that the most challenging points of a practical direct control implementation are the huge control set and the high-dimensional switched model when extending prediction horizon. The first class of solution methods originates from mathematical point of views, which applies complete and improved enumerative methods and searches for the minimal point of the cost function in an implicit way. The second class aims to reduce the size of the control sets by introducing physical limitation, which remarkably improve the algorithm efficiency but the prediction horizon still limits. The last class of solution method adopts the event-based concept and switches submodules referring to their event generators, which offers a prediction range to half of the period. As a conclusion, fast RCS method and event-based method offer the best possibility of practical implementation, which will be further analyzed in simulation and hardware.

9 Simulation and experimental verification of direct control

In this chapter the effectiveness and efficiency of MMCC direct control method are verified by simulation analysis and experimental study. The effectiveness is verified from tracking performance of AC external current, dynamic stability and balancing states of submodule voltages. The efficiency is examined from the perspectives of submodule switching frequency (power efficiency) and algorithm speed (computational efficiency). To evaluate direct control method, a conventional configuration with cascaded PI controllers and voltage-sorting modulation is selected as a benchmark.

9.1 Simulation verification

9.1.1 Simulation setup and parameters

A simulation model of the investigated one-phase MMC (MMCC-$\mathcal{K}_{3,1}$) is established in *Matlab/Simulink* with the help of toolbox *Simscape/SimPowerSystems*, whereby the nonlinear switching behavior of each MOSFET and influence of parasitic parameters are accounted. The parameters of the simulated system are given in Table 9.1, whose symbols correspond to Fig. 7.1. In order to reflect the inhomogeneity of submodule parameters and show the control performance, the submodule capacitance are selected with different values as listed in Table 9.2. These submodule parameters are scaled according to the real hardware system.

For the parameters in direct control, the system sampling period is set to 0.5ms, producing 40 switching decisions in one period and the weighting factors are fixed as

$$\begin{pmatrix} k_{ph} & k_{cir} & k_{\Sigma} & k_{\Delta} & k_{SM} \end{pmatrix} = \begin{pmatrix} 10 & 2 & 2 & 2 & 3 \end{pmatrix} \tag{9.1}$$

Table 9.1: Parameters of the simulated one-phase MMC

	number of branch SMs	N_{SM}	5
MMC branch	Branch inductance	L	2 mH
	Branch resistance	R	100 $m\Omega$
Submodule	Nominal voltage	U_{SM}	50 V
	Capacitance	C_{SM}	3.3 mF
SM capacitor	Equivalent series resistance	R_C	0.8 $m\Omega$
	self-inductance	L_C	0
	On-resistance	$R_{ds(on)}$	105 $m\Omega$
SM MOSFET	Internal inductance	L_{ds}	12 nH
	Body diode forward voltage	U_{fwd}	1.2 V
	DC voltage	u_{dc}	250 V
DC supply	DC-side resistance	R_{dc}	0
	DC-side inductance	L_{dc}	0
	Line frequency	f	50 Hz
AC load	AC-side resistance	R_{ac}	5 Ω
	AC-side inductance	L_{ac}	6 mH

Table 9.2: Unequal submodule capacitance in simulation setup

Submodule	Capacitance	Submodule	Capacitance
SM AP1	1.606 mF	SM AN1	1.608 mF
SM AP2	1.690 mF	SM AN2	1.613 mF
SM AP3	1.835 mF	SM AN3	1.605 mF
SM AP4	1.575 mF	SM AN4	1.662 mF
SM AP5	1.632 mF	SM AN5	1.641 mF

9.1.2 Direct control with exhaustive search

In order to show the effectiveness of direct control method, the first simulation study is based on the most straightforward solution scheme, e.g., exhaustive search for the total control set. The optimal switching combination is generated in each control period with a prediction horizon $N_{prd} = 1$, so that the external current i_{ph} is controlled to a sinusoidal waveform with an amplitude of 15 A. The simulation results are shown in Fig. 9.1. It can be seen that the controlled external current tracks the reference and the circulating current is also suppressed below 5A. With the help of one-step prediction based on individual submodule model, submodule voltages achieve an overall balanced distribution. The resulting averaging switching frequency for all the submodules is 476Hz, the degree of total harmonic distortion (THD) for terminal voltage is 18.5% and the degree for external current is 5.52%. The submodule voltage ripples are limited between 42V and 58V. The peak voltage difference is less than 3V (6% of the rated submodule voltage).

Direct control method based on exhaustive search cannot be regarded as a promising solution due to an inefficient search for the huge total control set. When the direct controller is adapted to two-step prediction, the simulation model runs extremely slow. It is already an impossible task to implement with any computational unit available in laboratory. The other two solution methods are therefore valuable to extend direct control to a possibly longer prediction horizon while still maintaining the practical time complexity and control effect.

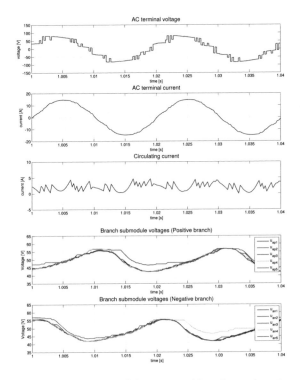

Figure 9.1: Simulation results of direct control based on exhaustive search

9.1.3 Direct control with fast RCS

Thanks to the high computational efficiency provided by fast RCS method and dynamic programming, the prediction horizon can easily be extended to 2. The simulation results are given in Fig. 9.2. The submodule switching frequency is reduced to 127Hz, voltage THD 25.9% and current THD 8.16%. Due to the principle limiting switching behavior, fast RCS method inherently achieves a low switching frequency at the expenses of higher THD, weaker suppression for circulating current and higher submodule voltage ripples. The voltage ripples reach maximal 61V and minimal 39V, and the maximal voltage difference for the submodule voltage peaks is as high as 14V (28% of the rated submodule voltage). The dynamical stability of submodule voltages is still ensured by fast RCS, but exact voltage balancing can hardly be achieved.

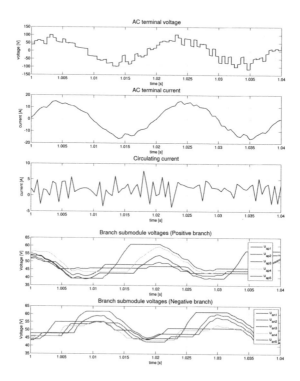

Figure 9.2: Simulation results of direct control based on fast RCS

With greatly reduced computational demand and switching frequency for fast RCS, the external current and voltage still achieve acceptable results compared with the exhaustive search. One side effect is that the stabilized submodule voltages are not equally distributed at each time instant, causing inhomogeneous voltage stress among the submodules and also therefore requiring a higher voltage margin for submodule design. In order to improve submodule voltage balancing, one possible solution is to extend the prediction horizon so that the submodule voltage ripples can be penalized along the horizon. However, as analyzed in Table 8.4, the computational amount for fast RCS with horizon 5 is almost equivalent to exhaustive search with horizon 1. Consider the simulation model with 40 control decisions in one period, prediction horizon 5 amounts to only $\frac{1}{8}$ period, which is not enough for a concrete prediction through the complete charging or discharging cycles.

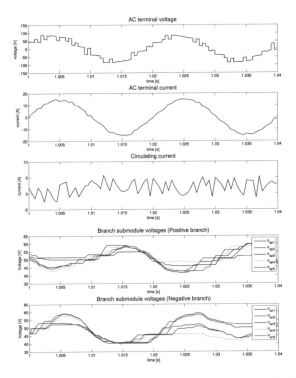

Figure 9.3: Simulation results of event-based method with conservative boundaries

9.1.4 Event-based direct control

The performance of event-based direct control, particularly the balancing state of sub-module voltages, depends on the upper and lower bounds. It is discussed in the following two settings.

Conservative bound setting
The upper and lower bounds are firstly set as 60V and 40V, allowing a voltage float $\pm 20\%$ of the rated submodule voltage. The simulation results for two periods are shown in Fig. 9.3. The external current is closely controlled to 15A with its THD value 5.51%. Meanwhile, the THD of the generated terminal voltage is 18.95%. The external current and voltage achieve comparable results to the solution scheme of exhaustive search, which is obviously better than fast RCS method. The averaging switching frequency is 305Hz. Although the submodule voltages are not balanced in a unified way through the entire period, their voltages comply firmly with the given upper and lower bounds.

Compressed bound setting
The estimated voltage ripple under 15A external current can be calculated as 43.7V and 56.5V. Thus, the upper and lower bounds are chosen as 43V and 57V by adding certain margin. The simulation results are shown in Fig. 9.4. The THDs of terminal voltage and current are 18.4% and 5.24%. By imposing compressed bound setting, the voltage peaks are well regulated and the maximal difference of submodule voltage peaks is around 2V. Compared with conservative bound setting, compressed setting increases the averaging submodule switching frequency to 326Hz. The event-based method with compressed bound setting will be chosen later for case comparison.

The event-based direct control method is a worthy improvement for solving MMCC direct control problem, which is shown by the following points: (1) A long prediction horizon covering the entire submodule charging or discharging cycles is achieved. It is implemented by each event generator in a distributed computing manner by means of look-up table; (2) The cost function to be evaluated is greatly simplified by eliminating 10 cost terms of submodule voltages from the original 14 ones, so that the problem dimension and the corresponding calculation complexity is greatly reduced; (3) A lower switching frequency is ensured while maintaining voltage boundaries and external current tracking.

9.1.5 Voltage-sorting method

In order for a fair comparison, cascaded PI controllers are employed for the simulation model to control AC external current, suppress circulating current and balance branch

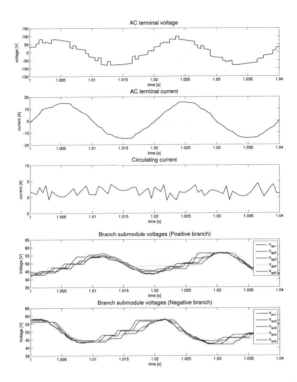

Figure 9.4: Simulation results of event-based method with compressed boundaries

energy. The generated branch voltage references from the controllers are modulated by phase-shifted modulation and then fed to a voltage-sorting-based switching block.

In this simulation study the MMC operation with two different submodule switching frequency, e.g., 350Hz and 130Hz, is investigated, which corresponds to the range of switching frequencies of event-based method and fast RCS method. The simulation results are given in Fig. 9.5. Fig. 9.5(a) shows the terminal voltage, external current, circulating current and submodule voltages when the switching frequency is 350Hz. The THD values are slightly better than event-based method (voltage THD 17.68% and current THD 3.17%). One feather is that all the branch submodule voltages have similar dynamic trajectories, meaning that their transient values maintain close to each other. However, due to different submodule capacitance and non-model-based one-step prediction, peak voltages of submodule capacitors may differ from each other. It can be seen that the maximal voltage difference at the peak values attains 3V.

Another simulation study alters the carrier frequency from 350Hz to 130Hz while the
rest simulation parameters remain the same. Compared with the simulation results
of 350Hz, the reduced submodule switching frequency leads to a worse current THD
(4.92%), amplified circulating current oscillation, higher submodule voltage peak (65V)
and larger voltage difference (maximal 17V). The consequences arising from voltage-
sorting method is more serious for practical medium-voltage application featured with a
small number of branch submodules, low switching frequency and possibly remarkable
difference in submodule parameters.

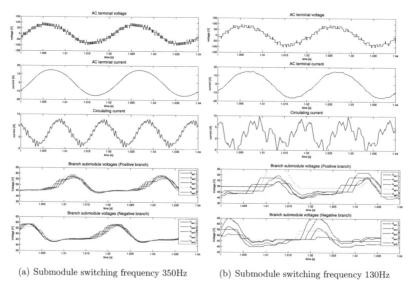

(a) Submodule switching frequency 350Hz   (b) Submodule switching frequency 130Hz

Figure 9.5: Simulation results of voltage-sorting method (350Hz and 130Hz)

9.1.6 Method comparison and time complex analysis

Switching performance analysis

The resulting submodule switching frequencies under different MMC control methods,
e.g., direct control method (exhaustive search, reduced control set and event-based
method) with $T_s = 0.5$ms and voltage-sorting method (350Hz and 130Hz), are summarized in Table 9.3 (FFC = Fundamental Frequency Component). Compared with

voltage-sorting method in the same switching frequency range, fast RCS offers a better
regulation of submodule voltage peaks, which suits for application scenarios requiring
extremely low switching frequency. Event-based method provides matchable voltage and
current THDs as voltage-sorting method with 350Hz, more importantly the regulation
of submodule voltage ripple and balancing of their peak voltages are superior.

Table 9.3: Performance comparison among different MMC control methods

		Direct control method ($T_s = 0.5$ms)			Voltage sorting method	
		Exhaustive search	Fast RCS	Event-based method	350Hz	130Hz
Averaging switching frequency [Hz]		476	127	326	350	130
Terminal voltage	THD	18.5%	25.9%	18.4%	17.7%	17.8%
	FFC [V]	78.61	79.54	79.53	80.96	81.49
External current	THD	5.52%	8.16%	5.24%	3.17%	4.92%
	FFC [A]	14.72	14.89	14.89	15.12	15.22
Submodule voltage range		[-16%,16%]	[-22%,22%]	[-14%,14%]	[-14%,18%]	[-20%,30%]
Maximal difference of submodule voltage peaks		6%	28%	4%	6%	34%

Time complexity comparison

The time complexity of voltage-sorting method and direct control method is compared
in this section. It is known that the commonly-used Bubble sort has worst-case and
average complexity both $O(N_{SM}^2)$, where N_{SM} is the number of items being sorted. The
time complexity of direct control method is listed in Table 9.4.

Event-based method has the least size of control set, which is determined by the law
of control set updating with a number of 9. Furthermore, it distributes the submodule
voltage balancing tasks ($2N_{SM}$ terms in cost function) to the event generators and then
has only four terms to be evaluated, e.g., external current, circulating current, sum
branch energy and branch energy difference. Therefore, time complexity of event-based
method remains constant regarding to an increased number of branch submodules N_{SM}.
On the other hand, event generators in event-based method are realized in parallel

Table 9.4: Time complexity from the solution schemes of direct control

		Direct control method		
		Exhaustive search	Reduced control set	Event-based method
Centralized computation	Control set size	$2^{2N_{SM}}$	$(N_{SM}+1)^2$	9
	Cost terms	$4+2N_{SM}$	$4+2N_{SM}$	4
	Time complexity	$O(N_{SM}4^{N_{SM}})$	$O(N_{SM}^3)$	$O(1)$
Distributed computation	Distributed calculation	No	No	Event generators and control set updating law
	Time complexity	-	-	$O(N_{SM})$
Computing increment with N_{prd}		Exponential	Exponential	Adaptive long horizon guaranteed

structures, which do not add time complexity. Although the maximal and minimal switching indices from the N_{SM} branch submodules are to be determined, it still remains as a linear time complexity of $O(N_{SM})$. Therefore, by analyzing the algorithms by Big O notation, it can be seen that event-based method is actually more efficient than the voltage-sorting method with quadratic time complexity and other solution schemes of direct control.

9.2 Experimental verification

To verify the proposed direct control methods, one-phase MMC is selected. The system parameters are listed here: $u_{dc} = 50$ V, submodule voltage rating $u_{SM,N} = 10$ V, $L = 0.5$ mH, $L_{ac} = 5.98$ mH and $R_{ac} = 14.83$ Ω. The line frequency is 50 Hz and the control period is set to 0.5 ms, meaning that in each period there exist 40 control decisions. The gate signals for two MOSFETs in one submodule are produced with a dead time 1.6 μs. The submodule capacitors are measured by a multimeter before they are mounted on board and afterward verified by the RC discharging circuit in the submodule board. Their values are listed from AP1 to AN5 as follow: 2.141 mF, 1.908 mF, 2.446 mF, 2.100 mF, 2.176 mF, 2.144 mF, 2.150 mF, 2.139 mF, 2.216 mF and 2.188 mF. Note that SM-AP2 has the minimal capacitance and SM-AP3 has the maximal capacitance, which will be specifically examined in different methods.

9.2.1 Experimental results of fast RCS

Steady-state performance when $I_{ph}^* = 1.5$ A

The measured terminal voltage, external current, two branch voltages, submodule voltages and switching signals for SM-AP2 and SM-AP3 are shown in Fig. 9.7. The submodule voltage ripples are +1.5 V and −1 V with respect to the reference value 10V. The voltage pressure is not appropriately shared among submodules and the maximal submodule voltage even occasionally reaches 12 V. Even though the dynamic stability of submodule voltages is still achieved, the unregulated upper bounds of submodule voltages may challenge the switch withstand voltage. The last two subfigures show the switching signals and the corresponding voltage variation for SM-AP2 and SM-AP3 (The DC offset 10V is eliminated from the red curve.). It can be seen that both of them have quite low switching frequency and SM-AP2 has a slightly higher switching frequency than SM-AP3.

Step response from $I_{ph}^* = 1$ A to 1.5 A

The control performance of step response is examined in Fig.9.6. The reference for external current is firstly set to 1 A and then jumps to 1.5 A. Without the delay introduced by modulation stage, a fast current tracking for the sinusoidal reference is realized for direct control method.

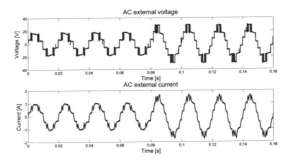

Figure 9.6: Step response under reduced control set from 1 A to 1.5 A

9.2.2 Experimental results of event-based method

In this part, the experimental results of event-based method are presented, as shown in Fig. 9.8. The current tracking is fulfilled, which is quite similar with the result of fast RCS method. The voltage spikes in the external voltage and branch voltages indicate the swap operation of submodule switching states occurring in control set updating. The prediction horizon for event-based method is extended to the entire charging and discharging cycle, so that the switching frequency rises and the voltage balancing performance is remarkably improved compared with fast RCS method. The submodule voltage ripple is between +1 V and −0.5 V. Meanwhile, it demands a much higher switching frequency.

9.2.3 Experimental results of voltage-sorting method

As a benchmark for experimental results of direct control methods, the classic voltage-sorting control method is selected. Fig. 9.9 shows the results when the external current reference is given for 1.5 A and switching frequency 800Hz. The last two subfigures in Fig. 9.9 show the submodule voltages and the corresponding gate signals. In order to guarantee a unified voltage distribution at each time instant, voltage-sorting method requires a frequent switching for the submodules so that the voltage ripple lies between +1 V and −0.5 V. It can also be seen that SM-AP2 with lower capacitance switches more frequently than SM-AP3 with higher capacitance.

9.3 Summary

In this chapter the switching performance and the algorithm efficiency of two proposed direct control schemes are examined and evaluated by comparing with the voltage-sorting method. The simulation results on a one-phase MMC with only 5 branch submodules show the feasibility of direct control method for external current control and internal submodule voltage balancing. Direct control method has the following advantages over voltage-sorting method: (1) fast and accurate control of external and circulating current, (2) dynamically stable submodule voltages with appropriate switching strategies and (3) submodule switching decision regarding to parametric inhomogeneity, given boundaries and future prediction. Fast RCS can inherently offer minimal switching frequency while ensuring improved harmonic performance and submodule voltage regulation, which is suitable for medium-voltage application. Although direct control method could bring a high time complexity for a large number of branch submodules and an increased prediction horizon, event-based method is always capable of providing a constrained time complexity while guaranteeing the adaptively long horizon.

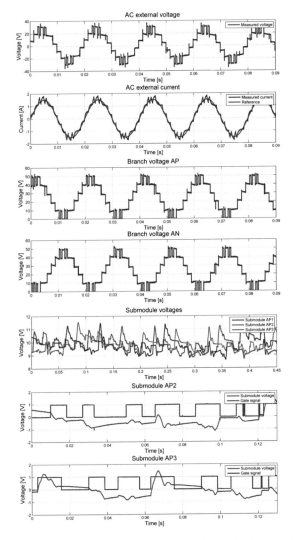

Figure 9.7: Experimental results of fast RCS method for 1.5A

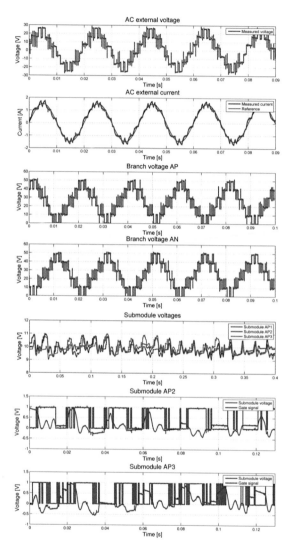

Figure 9.8: Experimental results of event-based method for 1.5A

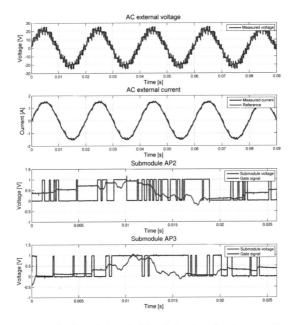

Figure 9.9: Internal submodule switching behavior of voltage-sorting method

10 Conclusion

The novel MMCC system is featured with a high quantity of coupled system variables, e.g., converter currents, branch energies and distributed capacitor voltages, and abundant discrete control inputs, e.g., submodule switching states. These variables are required to be fully modelled and controlled for a stable and optimal system operation, which is still a fundamental problem of MMCC research. To tackle this challenge, this thesis has presented two frameworks of modeling and analyzing MMCCs.

The first framework in Part I presents a comprehensive classification of MMCC topologies, analyzes them by replacing converter branches with continuous controllable voltage sources and develops a unified modeling method for MMCC systems, aiming for a general understand of MMCC in the context of continuous system theory. The following work has been implemented:

1. Based on the continuous equivalent circuit of MMCC, a unified framework of modeling MMCC systems has been proposed by dividing into four decomposed parts, which are external current modeling, internal current modeling, external power modeling and internal branch energy balancing. For future practical implementation, this framework can be efficiently supported by a proper algebraic modeling tool aiming for automated model generation.

2. Circulating current that widely exists in MMCCs has been defined from physical aspect and mathematically formulated. A layered internal branch energy balancing strategy has been proposed by using the specific frequency components in circulating currents to realize an equal energy distribution among MMCC branches.

3. A set of MMCC control schemes based on multivariable control theory and optimal principle have been proposed to handle the MMCC control issue. The multivariable optimal control schemes, e.g., state feedback control using pole placement, LQR, nonlinear feedback control based on SDRE and periodic discrete LQR, have been presented for different MMCC application scenarios.

4. In order to validate the modeling method and multivariable control, simulation studies for three MMCC systems have been made and experimental verification based on a laboratory MMC prototype has been given.

The second framework aims to generate submodule switching states for realizing external control and internal energy balancing by using the concept of direct torque control. It develops an explicit relation between submodule switching states and MMCC system states, which analyzes MMCC by preserving its essential feature of discrete switched system. The involved work is listed here:

1. A direct switched model has been developed for MMCC by taking one-phase MMC (HB-$\mathcal{K}_{2,1}$) as example. Based on the direct switched model, the problem of MMCC direct control has been generally formulated into multi-step discrete optimization.

2. Reduced control set (RCS) methods have been introduced to increase online calculation rate and extend the prediction horizon. Fast RCS method that follows the principle of restricting switching behavior and combines current prediction based on delta-input formulation has been developed for practical implementation and enhanced dynamic performance.

3. Event-based direct control method has been developed to achieve the minimal computation effort and guarantee an adaptively long prediction horizon by applying event-triggered mechanism for submodule voltage regulation.

4. By taking the conventional voltage sorting method as a benchmark, fast RCS and event-based direct control methods have been validated by a series simulation and experimental studies. Under the premise of the same switching frequency as the voltage sorting method, direct control methods have achieved comparable harmonic performance and improved submodule voltage regulation.

Appendices

A Relation between branch energy and branch submodule voltages

Each branch with cascaded submodules is substituted by one CVCS. The branch power and branch energy is related to the power and energy of the CVCS. The instantaneous branch power can be written as

$$p_{j,k}(t) = u_{j,k}(t)i_{j,k}(t) \tag{A.1}$$

Consider the l-th submodule ($l = 1, 2, ..., N_{SM}$) in Branch $E_{j,k}$, the relation between the branch current $i_{j,k}(t)$ and the capacitor voltage $u_{c,jk,l}(t)$ is given with the help of its switching state $g_{jk,l}(t)$ as

$$g_{jk,l}(t)i_{j,k}(t) = -C\frac{du_{c,jk,l}(t)}{dt} \tag{A.2}$$

It shows that the submodule voltage $u_{c,jk,l}(t)$ changes depending on the branch current $i_{j,k}(t)$ and the switching state $g_{jk,l}(t)$ of this submodule. Since the branch voltage $u_{j,k}(t)$ is

$$u_{j,k}(t) = \sum_{l=1}^{N_{SM}} g_{jk,l}(t)u_{c,jk,l}(t) \tag{A.3}$$

By summarizing the equation for Branch $E_{j,k}$ in (A.1), (A.2) and (A.3), the instantaneous branch power $p_{j,k}(t)$ can be derived to the submodule layer as:

$$
\begin{aligned}
p_{j,k}(t) = u_{j,k}(t)i_{j,k}(t) &= \sum_{l=1}^{N_{SM}} g_{jk,l}(t)u_{c,jk,l}(t)i_{j,k}(t) \\
&= \sum_{l=1}^{N_{SM}} u_{c,jk,l}(t)(g_{jk,l}(t)i_{j,k}(t)) \\
&= -\sum_{l=1}^{N_{SM}} u_{c,jk,l}(t)C\frac{du_{c,jk,l}(t)}{dt} \\
&= -\sum_{l=1}^{N_{SM}} \frac{1}{2}C\frac{du_{c,jk,l}^2(t)}{dt} \\
&= -\frac{d}{dt}\sum_{l=1}^{N_{SM}}\left(\frac{1}{2}Cu_{c,jk,l}^2(t)\right)
\end{aligned}
\tag{A.4}
$$

An integration of the above expression gives the branch energy as

$$w_{j,k}(t) = \sum_{l=1}^{N_{SM}}\left(\frac{1}{2}Cu_{c,jk,l}^2(t)\right) \tag{A.5}$$

The expression (A.5) illustrates that the branch energy is directly related to its sum submodule energy e.g. submodule voltages, even though the SMs are switched on and off during run-time.

According to the defined positive directions of the branch voltage and the branch current in this thesis, the branch energy $w_{j,k}(t)$ is related by the branch power as:

$$\dot{w}_{j,k}(t) = -p_{j,k}(t) = -u_{j,k}(t)i_{j,k}(t) \tag{A.6}$$

When the branch voltage acts as a power supply with a positive branch current, it will release energy and the energy stored in the capacitors is reduced. It explains the minus sign in (A.6).

B Circulating currents in MMCC

The circulating current expressions are directly given here for MMCC-$\mathcal{K}_{M,N}$ and -$\mathcal{C}_{M+N}$:

MMCC-$\mathcal{K}_{M,N}$

For MMCC-$\mathcal{K}_{2,2}$, the only circulating current is

$$i_z = \frac{1}{4}(i_{1,1} + i_{1,2} + i_{2,1} + i_{2,2}) \tag{B.1}$$

For MMCC-$\mathcal{K}_{3,2}$, the both circulating currents are

$$\begin{cases} i_{z,1} = \frac{1}{6}i_{1,1} - \frac{1}{2}i_{1,2} - \frac{1}{3}i_{2,1} - \frac{1}{3}i_{3,1} \\ i_{z,2} = -\frac{1}{3}i_{1,1} + \frac{1}{6}i_{2,1} - \frac{1}{2}i_{2,2} - \frac{1}{3}i_{3,1} \end{cases} \tag{B.2}$$

For MMCC-$\mathcal{K}_{3,3}$, the four circulating currents can also be derived as

$$\begin{cases} i_{z,1} &= \frac{1}{3}i_{1,1} - \frac{1}{3}i_{1,2} - \frac{1}{3}i_{1,3} - \frac{1}{3}i_{2,1} - \frac{1}{3}i_{3,1} \\ i_{z,2} &= -\frac{1}{3}i_{1,1} + \frac{1}{3}i_{1,2} - \frac{1}{3}i_{1,3} - \frac{1}{3}i_{2,2} - \frac{1}{3}i_{3,2} \\ i_{z,3} &= -\frac{1}{3}i_{1,1} + \frac{1}{3}i_{2,1} - \frac{1}{3}i_{2,2} - \frac{1}{3}i_{2,3} - \frac{1}{3}i_{3,1} \\ i_{z,4} &= -\frac{1}{3}i_{1,2} - \frac{1}{3}i_{2,1} + \frac{1}{3}i_{2,2} - \frac{1}{3}i_{2,3} - \frac{1}{3}i_{3,2} \end{cases} \tag{B.3}$$

MMCC-$\mathcal{C}_{M+N}$

For MMCC-$\mathcal{C}_3$, the only circulating current is calculated as

$$i_z = \frac{1}{3}(i_1 + i_2 + i_3) \tag{B.4}$$

For MMCC-$\mathcal{C}_6$,

$$i_z = \frac{1}{6}(i_{1,1} + i_{2,1} + i_{2,3} + i_{3,3} + i_{3,2} + i_{1,2}) \tag{B.5}$$

Table B.1 summarizes the quantity of the independent circulating currents for MMCC topologies, and the selected circulating currents $\boldsymbol{i}_z$.

Table B.1: Circulating currents (CCs) in the MMCC topologies

Graph type	Quantity of branch N_{br}	Quantity of terminal N_{tm}	Quantity of independent CCs	CC symbols
$\mathcal{K}_{2,1}$	2	3	0	$\emptyset$
$\mathcal{K}_{2,2}$	4	4	1	$i_z = i_{z,11}$
$\mathcal{K}_{3,1}$	3	4	0	$\emptyset$
$\mathcal{K}_{3,2}$	6	5	2	$i_{z1} = i_{z,11},\, i_{z2} = i_{z,21}.$
$\mathcal{K}_{3,3}$	9	6	4	$i_{z1} = i_{z,11},\, i_{z2} = i_{z,12},$ $i_{z3} = i_{z,21},\, i_{z4} = i_{z,22}.$
$\mathcal{K}_{M,N}$	$M \times N$	$M + N$	$N_{br} - N_{tm} + 1$	$i_{z1},\, ...,\, i_{z(N_{br}-N_{tm}+1)}.$
$\mathcal{C}_3$	3	3	1	i_z
$\mathcal{C}_6$	6	6	1	i_z

Bibliography

[1] B.K. Bose. Power electronics and motor drives recent progress and perspective. *Industrial Electronics, IEEE Transactions on*, 56:581–588, 2009.

[2] Hidetoshi Umida. Power electronics technology trends and prospects. *Fuji Electric Journal*, 75:1–5, 2002.

[3] A. Akdag. Soa in high power semiconductors. In *Industry Applications Conference, 2006. 41st IAS Annual Meeting. Conference Record of the 2006 IEEE*, 2006.

[4] B. Jayant Baliga. Trends in power semiconductor devices. *Electron Devices, IEEE Transactions on*, 43:1717–1731, 1996.

[5] J. Vobecky. Future trends in high power devices. In *Microelectronics Proceedings (MIEL), 2010 27th International Conference on*, 2010.

[6] J. Rodriguez, L.G. Franquelo, S. Kouro, J.I. Leon, R.C. Portillo, M.A.M. Prats, and M.A. Perez. Multilevel converters: An enabling technology for high-power applications. *Proceedings of the IEEE*, 97:1786–1817, 2009.

[7] M. Malinowski, K. Gopakumar, J. Rodriguez, and M.A. Perez. A survey on cascaded multilevel inverters. *Industrial Electronics, IEEE Transactions on*, 57:2197–2206, 2010.

[8] S. Kouro, M. Malinowski, K. Gopakumar, J. Pou, L.G. Franquelo, Bin Wu, J. Rodriguez, M.A. Perez, and J.I. Leon. Recent advances and industrial applications of multilevel converters. *Industrial Electronics, IEEE Transactions on*, 57:2553–2580, 2010.

[9] A. Nabae, I. Takahashi, and H. Akagi. A new neutral-point-clamped pwm inverter. *Industry Applications, IEEE Transactions on*, IA-17:518–523, 1981.

[10] L.M. Tolbert, Fang Zheng Peng, and T.G. Habetler. A multilevel converter-based universal power conditioner. *Industry Applications, IEEE Transactions on*, 36:596–603, 2000.

[11] H. Akagi. New trends in medium-voltage power converters and motor drives. In

2011 IEEE International Symposium on Industrial Electronics, 2011.

[12] T. A. Meynard and H. Foch. Multi-level conversion: high voltage choppers and voltage-source inverters. In *Power Electronics Specialists Conference, 1992. PESC '92 Record., 23rd Annual IEEE*, 1992.

[13] Jih-Sheng Lai and Fang Zheng Peng. Multilevel converters-a new breed of power converters. *Industry Applications, IEEE Transactions on*, 32:509–517, 1996.

[14] S. Debnath, Jiangchao Qin, B. Bahrani, M. Saeedifard, and P. Barbosa. Operation, control, and applications of the modular multilevel converter: A review. *Power Electronics, IEEE Transactions on*, 30:37–53, 2015.

[15] H. Akagi. Classification, terminology, and application of the modular multilevel cascade converter (mmcc). *Power Electronics, IEEE Transactions on*, 26:3119–3130, Nov. 2011.

[16] R. Marquardt. Modular multilevel converter: An universal concept for hvdc-networks and extended dc-bus-applications. In *Power Electronics Conference (IPEC), 2010 International*, 2010.

[17] M. Hagiwara, R. Maeda, and H. Akagi. Negative-sequence reactive-power control by a pwm statcom based on a modular multilevel cascade converter (mmcc-sdbc). *Industry Applications, IEEE Transactions on*, 48:720–729, 2012.

[18] Jun Mei, Bailu Xiao, Ke Shen, L.M. Tolbert, and Jian Yong Zheng. Modular multilevel inverter with new modulation method and its application to photovoltaic grid-connected generator. *Power Electronics, IEEE Transactions on*, 28:5063–5073, 2013.

[19] J. Peralta, H. Saad, S. Dennetiere, J. Mahseredjian, and S. Nguefeu. Detailed and averaged models for a 401-level mmc-hvdc system. *Power Delivery, IEEE Transactions on*, 27:1501–1508, 2012.

[20] R. Marquardt, A. Lesnicar, and J. Hildinger. Modulares stromrichterkonzept für netzkupplungsanwendungen bei hohen spannungen. In *ETG-Fachtagung*, 2002.

[21] M. Glinka. Prototype of multiphase modular-multilevel-converter with 2 mw power rating and 17-level-output-voltage. In *Power Electronics Specialists Conference, 2004. PESC 04. 2004 IEEE 35th Annual*, 2004.

[22] *Siemens AG. HVDC PLUS. Refer to: http://www.energy.siemens.com.*

[23] Noman Ahmed, Staffan Norrga, H.-P. Nee, A. Haider, D. Van Hertem, Lidong Zhang, and Lennart Harnefors. Hvdc supergrids with modular multilevel converters

- the power transmission backbone of the future. In *Systems, Signals and Devices (SSD), 2012 9th International Multi-Conference on*, 2012.

[24] D. Schmitt, Y. Wang, T. Weyh, and R. Marquardt. Dc-side fault current management in extended multiterminal-hvdc-grids. In *Systems, Signals and Devices (SSD), 2012 9th International Multi-Conference on*, 2012.

[25] M.M.C. Merlin, T.C. Green, P.D. Mitcheson, D.R. Trainer, R. Critchley, W. Crookes, and F. Hassan. The alternate arm converter: A new hybrid multilevel converter with dc-fault blocking capability. *Power Delivery, IEEE Transactions on*, 29:310–317, 2014.

[26] M.A. Perez, R. Lizana, C. Azocar, J. Rodriguez, and Bin Wu. Modular multilevel cascaded converter based on current source h-bridges cells. In *IECON 2012 - 38th Annual Conference on IEEE Industrial Electronics Society*, 2012.

[27] L. Baruschka and A. Mertens. Comparison of cascaded h-bridge and modular multilevel converters for bess application. In *Energy Conversion Congress and Exposition (ECCE), 2011 IEEE*, 2011.

[28] E. Solas, G. Abad, J.A. Barrena, S. Aurtenetxea, A. Carcar, and L. Zajac. Modular multilevel converter with different submodule concepts part i: Capacitor voltage balancing method. *Industrial Electronics, IEEE Transactions on*, 60:4525–4535, 2013.

[29] Kui Wang, Yongdong Li, Zedong Zheng, and Lie Xu. Voltage balancing and fluctuation-suppression methods of floating capacitors in a new modular multilevel converter. *Industrial Electronics, IEEE Transactions on*, 60:1943–1954, 2013.

[30] M. Pereira, D. Retzmann, J. Lottes, M. Wiesinger, and G. Wong. Svc plus: An mmc statcom for network and grid access applications. In *PowerTech, 2011 IEEE Trondheim*, 2011.

[31] *Siemens AG. SVC PLUS. Refer to: http://www.energy.siemens.com.*

[32] C. Oates. A methodology for developing 'chainlink' converters. In *Power Electronics and Applications, 2009. EPE '09. 13th European Conference on*, 2009.

[33] Felix Kammerer, Dennis Braeckle, Mario Gommeringer, Mathias Schnarrenberger, and Michael Braun. Operating performance of the modular multilevel matrix converter in drive applications. In *PCIM Europe 2015 Proceedings of*, 2015.

[34] L. Baruschka and A. Mertens. A new 3-phase direct modular multilevel converter. In *Power Electronics and Applications (EPE 2011), Proceedings of the 2011-14th*

European Conference on, 2011.

[35] L. Baruschka and A. Mertens. A new 3-phase ac/ac modular multilevel converter with six branches in hexagonal configuration. *Industry Applications, IEEE Transactions on*, PP:1–1, 2013.

[36] E. C. dos Santos and E. R. C. da Silva. Power block geometry applied to the building of power electronics converters. *IEEE Transactions on Education*, 56:191–198, 2013.

[37] Bjorn Jacobson, Patrik Karlsson, Gunnar Asplund, Lennart Harnefors, and Tomas Jonsson. Vsc-hvdc transmission with cascaded two-level converters. In *International Council on Large Electric Systems (CIGRE)*, 2010.

[38] M. Glinka and R. Marquardt. A new single phase ac/ac-multilevel converter for traction vehicles operating on ac line voltage. In *European Conference on Power Electronics and Applications*, 2003.

[39] Yun Wan, Steven Liu, and Jianguo Jiang. Generalized analytical methods and current-energy control design for modular multilevel cascade converter. *IET Power Electronics*, 6:495–504, 2013.

[40] M. Hagiwara and H. Akagi. Control and experiment of pulsewidth-modulated modular multilevel converters. *Power Electronics, IEEE Transactions on*, 24:1737–1746, 2009.

[41] Lennart Angquist, A. Antonopoulos, D. Siemaszko, K. Ilves, M. Vasiladiotis, and H-P Nee. Open-loop control of modular multilevel converters using estimation of stored energy. *Industry Applications, IEEE Transactions on*, 47:2516–2524, 2011.

[42] D.C. Ludois, J.K. Reed, and G. Venkataramanan. Hierarchical control of bridge-of-bridge multilevel power converters. *Industrial Electronics, IEEE Transactions on*, 57:2679–2690, 2010.

[43] A. Antonopoulos, Lennart Angquist, and H-P Nee. On dynamics and voltage control of the modular multilevel converter. In *Power Electronics and Applications, 2009. EPE '09. 13th European Conference on*, 2009.

[44] Gene F. Franklin, J. D. Powell, and Michael Workmann. *Digital Control of Dynamic Systems*. Addison Wesley Longman, Inc., 1998.

[45] Gene F. Franklin, J. David Powell, and Abbas Emami Naeini. *Feedback Control of Dynamic Systems*. Prentice-Hall, Upper Saddle River, New Jersey, 2006.

[46] P. P. Albertos and A. Sala. *Multivariable control of systems: an engineering approach*. Springer-Verlag London, 2004.

[47] D. Subbaram Naidu. *Optimal Control Systems*. CRC Press, 2002.

[48] H.T. Banks, B.M. Lewis, and H.T. Tran. Nonlinear feedback controllers and compensators: a state-dependent riccati equation approach. *Computational Optimization and Applications*, 37:177–218, 2007.

[49] J. Pittner, N.S. Samaras, and M.A. Simaan. A new strategy for optimal control of continuous tandem cold metal rolling. *Industry Applications, IEEE Transactions on*, 46:703–711, 2010.

[50] H. T. Banks, R. C H Del Rosario, and R.C. Smith. Reduced-order model feedback control design: numerical implementation in a thin shell model. *Automatic Control, IEEE Transactions on*, 45:1312–1324, 2000.

[51] E.B. Erdem and A.G. Alleyne. Experimental real-time sdre control of an underactuated robot. In *Decision and Control, 2001. Proceedings of the 40th IEEE Conference on*, 2001.

[52] T. Fujimoto, F. Tabuchi, and T. Yokoyama. Digital control of single phase pwm inverter using sdre approach. In *Power Electronics Conference (IPEC), 2010 International*, 2010.

[53] Ton Duc Do, Han Ho Choi, and Jin-Woo Jung. Sdre-based near optimal control system design for pm synchronous motor. *Industrial Electronics, IEEE Transactions on*, 59:4063–4074, 2012.

[54] James R. Cloutier. State-dependent riccati equation techniques: an overview. In *American Control Conference, 1997. Proceedings of the 1997*, 1997.

[55] Tayfun Cimen. State-dependent riccati equation (sdre) control: A survey. In *Proceedings of the 17th IFAC World Congress*, 2008.

[56] T. Yucelen, A.S. Sadahalli, and F. Pourboghrat. Online solution of state dependent riccati equation for nonlinear system stabilization. In *American Control Conference (ACC), 2010*, 2010.

[57] T. Yucelen, P.V. Medagam, and F. Pourboghrat. Nonlinear quadratic optimal control for cascaded multilevel static compensators. In *Power Symposium, 2007. NAPS '07. 39th North American*, 2007.

[58] Chun-Liang Lin, Chi-Chih Lai, and Teng-Hsien Huang. A neural network for linear matrix inequality problems. *Neural Networks, IEEE Transactions on*, 11:1078–1092, 2000.

[59] P. Munch, D. Gorges, M. Izak, and S. Liu. Integrated current control, energy

control and energy balancing of modular multilevel converters. In *IECON 2010 - 36th Annual Conference on IEEE Industrial Electronics Society*, 2010.

[60] Sergio Bittanti and Patrizio Colaneri. *Periodic Systems: Filtering and Control.* Springer London, 2009.

[61] Daniel Görges. *Optimal Control and Scheduling of Switched Systems.* PhD thesis, Control Systems Institute, Kaiserslautern, 2011.

[62] Philipp Münch. *Konzeption und Entwurf integrierter Regelungen für Modulare Multilevel Umrichter.* Logos Verlag Berlin, 2011.

[63] G.S. Konstantinou and V.G. Agelidis. Performance evaluation of half-bridge cascaded multilevel converters operated with multicarrier sinusoidal pwm techniques. In *Industrial Electronics and Applications, 2009. ICIEA 2009. 4th IEEE Conference on*, 2009.

[64] A. Lesnicar and R. Marquardt. An innovative modular multilevel converter topology suitable for a wide power range. In *Power Tech Conference Proceedings, 2003 IEEE Bologna*, 2003.

[65] L.G. Franquelo, J. Rodriguez, J.I. Leon, S. Kouro, R. Portillo, and M.A.M. Prats. The age of multilevel converters arrives. *Industrial Electronics Magazine, IEEE*, 2:28–39, 2008.

[66] Qingrui Tu and Zheng Xu. Impact of sampling frequency on harmonic distortion for modular multilevel converter. *Power Delivery, IEEE Transactions on*, 26:298–306, 2011.

[67] Jiangchao Qin and M. Saeedifard. Predictive control of a modular multilevel converter for a back-to-back hvdc system. *Power Delivery, IEEE Transactions on*, 27:1538–1547, 2012.

[68] M. Glinka and R. Marquardt. A new ac/ac multilevel converter family. *Industrial Electronics, IEEE Transactions on*, 52:662–669, 2005.

[69] Qingrui Tu, Zheng Xu, and Lie Xu. Reduced switching-frequency modulation and circulating current suppression for modular multilevel converters. *Power Delivery, IEEE Transactions on*, 26:2009–2017, 2011.

[70] T. Nagel, J. Ismail, Y. Wan, and S. Liu. Design of mvdc power transmission for a grid connected agricultural machine. In *2016 IEEE Vehicle Power and Propulsion Conference (VPPC)*, 2016.

[71] A. Draou and S. Mieee. A space vector modulation based three-level pwm rectifier

under simple sliding mode control strategy. *Energy and Power Engineering*, 5:28–35, 2013.

[72] C Wessels, S. Grunau, and F. W. Fuchs. Current injection targets for a statcom under unbalanced grid voltage condition and the impact on the pcc voltage. In *EPE joint wind energy and TD chapters seminar*, 2011.

[73] T. Geyer and D. E. Quevedo. Multistep direct model predictive control for power electronics part 2: Analysis. In *2013 IEEE Energy Conversion Congress and Exposition*, 2013.

[74] R.E. Bellman and S.E. Dreyfus. *Applied Dynamic Programming*. Princeton University Press, 1962.

[75] Ulrich Hoffmann. *Entwurf und Erprobung einer Adapativen Zweipunktregelung*. PhD thesis, RWTH Aachen, 1984.

[76] Stephen P. Boyd. Convex optimization ii - branch and bound methods. Lecture slides.

[77] S. Boyd and L. Vandenberghe. localization and cutting-plane methods. September 18, 2003.

[78] R. D. C. Monteiro. First- and second-order methods for semidefinite programming. *Mathematical Programming*, 97:209–244, 2003.

[79] Johannes Terno. *Numerische Verfahren der diskreten Optimierung*. Teubner, 1981.

[80] Z. Li, F. Gao, F. Xu, X. Ma, Z. Chu, P. Wang, R. Gou, and Y. Li. Power module capacitor voltage balancing method for a 350-kv/1000-mw modular multilevel converter. *IEEE Transactions on Power Electronics*, 31:3977–3984, 2016.

Zusammenfassung

In den letzten Jahren hat die Familie der Multilevel-Umrichtern auch die so sogenannten kaskadierten modularen Multilevel-Umrichter (Modular Multilevel Cascaded Converter, MMCC), mit modularer Realisierung und flexibler Spannungspegelerweiterung, in der wissenschaftlichen Forschung und in der industriellen Anwendung zunehmend an Bedeutung gewonnen. Im Vergleich zu den herkömmlichen Multilevel-Topologien, wie Diode-Clamped-Umrichter, Flying-Capacitor-Umrichter und Kaskadierten-H-Brücken-Umrichter, bietet der MMCC vorteilhafte Eigenschaften in Bezug auf die modulare Implementierung, einfache Erweiterbarkeit der Spannungspegels, Skalierbarkeit für verschiedene Spannungen und Leistungen. Zu den weiteren Vorteilen gehört die wenigen Oberschwingungen und die hohe Energieumwandlungseffizienz. Die MMCC-Familie verkörpert eine vielversprechende Lösung für die Zukunft der Hochspannungs- und Hochleistungsanwendung. Dazu gehören Static Synchronous Compensator (STATCOM) zur Kompensation von Blindleistung, Batterien-Energiespeicherungssysteme für aktive Leistungssteuerung, Hochspannungs-Gleichstrom-Übertragung-Technologien, AC/AC-Umwandlung für Motorantriebe sowie andere Anwendungen im Schienenverkehrsnetzwerk.

In den bisherigen Publikationen werden die analytischen Methoden anhand einer einzelnen Topologie aus der MMCC-Familie untersucht. Dabei werden die inneren Zusammenhänge vernachlässigt. Es gibt jedoch eine viel Zahl von MMCC, die sich im Wesentlichen anhand der topologischen Struktur unterscheiden. Die Art der einzelnen Submodulen (Halb- oder Vollbrücken) sowie die Struktur (Verdopplung, Verdreifachung) ergeben eine viel Zahl an Konfigurationsmöglichkeiten für unterschiedlich funktionierende MMCCs. Trotz Unterschiede finden sich Gemeinschaften unterhalb der MMCCs, die in der Betrachtung ein Bild zu den operationalen Prinzipien wiedergeben. Im Rahmen dieser Dissertation werden die unterschiedlichen MMCCs systematisch näher untersucht und eine bessere Übersicht gegeben.

Diese Dissertation besteht aus zwei Teilen. Der erste Teil (Kapitel 2-6) fokussiert auf eine einheitliche Modellierungsmethode für MMCCs im Sinne eines kontinuierlichen Systems. Der zweite Teil (Kapitel 7-9) beschäftigt sich mit der sogenannten direkten Regelung anhand eines geschalteten Hybridsystems.

Kapital 1 enthält einen Überblick über die unterschiedlichen Multilevel-Technologien und die bestehenden MMCC-Systeme. Hierzu werden die Motivation zu einer allgemeinen Modellierungsmethode für MMCC und die direkte Regelung erläutert und im Anschluss wird der eigene Beitrag dieser Dissertation hervorgehoben.

Im Kapital 2 werden die in der Arbeit angewandten Konzepte für MMCC eingeführt und tiefe Einblicke in die unterschiedlichen MMCCs gewährt. Im Weiteren werden zur Klassifizierung der MMCC zwei Methoden aus der aktuellen Forschung vorstellt, die jedoch keine allgemeine Klassifizierung anbieten. Deshalb wird eine weitere Methode vorgestellt, die eine umfassende Klassifikation in Betracht zieht und auf der Graphentheorie basiert ist. Die entstandene Klassifizierung liefert eine eindeutige Bezeichnung und eine strukturelle Simplifikation für alle MMCC. Die auf Graphentheorie basierende Analyse bezieht Informationen über die Graphen der einzelnen MMCC-Topologien und verwendet diese als Referenzen für die Namensfindung und Symbolisierung.

Im Kapital 3 wird die vorgestellte Methode zur allgemeinen Modellierung vorgestellt, die unabhängig von den Topologien ist und deshalb für alle MMCC geeignet ist. Hierzu werden die MMCC in Spannung- und Energiestufen untergeteilt. Solch eine Aufteilung ist vor allem für die Regelung von großer Bedeutung, da unter den Spannung- und Energiestufen Netzströmen und Kreisströmen sich wieder finden. Die Wirk- und Blindleistungen werden durch die Netzströme geregelt. Dazu werden die Energien der einzelnen Kondensatoren bzw. der Zweigen durch die Regelung der Kreisströme balanciert. Weitere Energiebalancierungsmethode wird entwickelt, um die Energieverteilung unter den MMCC-Zweigen auszugleichen. Die allgemeindefinierten Kreisströme spielen bei der Unterstützung des Energieaustauschs zwischen den Zweigen eine wesentliche Rolle und tragen letztendlich zu einem ausgeglichenen Ziel bei. Der MMCC wird vollständig im Zustandsraum modelliert, sodass die Voraussetzungen für das Erreichen der oben genannten Regelungsaufgaben erfüllt werden.

Kapitel 4 befasst sich mit der Zustandsraumdarstellung und der multivariablen optimalen Regelung für MMCC. Unter Rücksicht der Nichtlinearität und zeitvarianten, Eigenschaften wurde der MMCC als ein Mehrgrößensystem mit vielen Kopplungseffekten Schritt für Schritt ausgearbeitet. Bekannte Entwurfsverfahren wie LQR oder Polvorgabe werden durchgeführt (Abschnitt 4.1 und 4.2). Abschnitt 4.3 beschäftigt sich mit der Nichtlinearität im MMCC-Modell, die nichtlineare quadratische Regler in Rahmen einer zustandssabhängigen Riccati-Gleichung direkt bearbeitet. In Abschnitt 4.4 wird die bilineare Eigenschaft von MMCC-Modellen durch die Linearisierung im Arbeitspunkt analysiert und eliminiert. Das ursprüngliche lineare zeitvariable (LTV) Modell wird als p-periodisches System mit Hyper-Periode modelliert. Im weiteren Schritt wird das System als ein periodisches diskretes System mit p linearen zeitinvarianten (LTI) Teilsystemen für jeden Schritt der Hyper-Periode diskretisiert. Dementsprechend wird ein periodischer LQR mit Vorsteuerung der Führungsgröße und Störungsgröße vorgestellt, um das resultierende p-periodische diskrete LTI-System zu regeln.

Kapital 5 enthält drei Simulationsstudien über MMC zur Übertragung von Wirkleistung, modularen kaskadierten STATCOM zur Kompensation von Blindleistung und modularen multilevel Matrix Umrichter zur Stromnetzverbindung. Anschließend wird die allgemeine Modellierungsmethode und der einheitliche Regelungsentwurf angewendet. Die Simulationsergebnisse unter realitätstreuen Bedingungen zeigen, dass der Regler eine bessere Performance erreichen kann.

In Kapitel 6 wird das im Labor implementierte MMC-System erläutert. Weiterhin werden experimentelle Ergebnisse auf Basis der eingeführten Modellierungsmethode vorgestellt und diskutiert.

Anschließend wird in Kapitel 7 auf die Modellierungsmethode eingegangen und die direkte Regelung von MMCC anhand der Zustandsraumdarstellung eingeführt. Bei der Modellierung wird explizit der Zusammenhang zwischen dem Schalten der MMC Submodule und der Systemzustände berücksichtigt. Beispielhaft wird hierzu die Vorgehensweise an einem einphasigen MMC ausführlich beschrieben. Basierend auf diesem Modell, wird die Regelung als diskretes Mehrschritt-Optimierungsproblem mit integrierter Kostenfunktion entwickelt. Hierbei entstehen eine Vielzahl Schaltmöglichkeiten.

Im Kapital 8 stehen die Lösungsschemen für direkte Regelungsmethode im Vordergrund. Die Implizierte Methode findet die optimale Lösung durch numerische Suchalgorithmen, z.B. den mühsamen Aufsuchungsalgorithmus (exhaustive search) und die dynamische Programmierung. Bei großen Systemen mit vielen Zustanden wird die numerische Suche noch mehr zeitaufwendig, sodass eine echtzeitfähige Lösung gar nicht möglich wird. Eine andere Perspektive bietet die Methoden der reduzierten Stellgrößenmengen. Die Reduktion basiert zum einen auf einer Klassifizierung der möglichen Schaltredundanz aber auch auf einer Beschränkung der Schaltfrequenz. Die ereignisbasierte direkte Regelungsmethode weist jedem Submodul einen Ereignisgenerator zu. Der Ereignisgenerator bewertet das Schaltverhältnis basiert auf den prädizierten Zweigströmen und den Submodulinternen Parameter. Daraus wird eine optimale Schaltsequenz generiert.

In Kapitel 9 werden die Simulationsergebnisse zur direkten Regelung der einphasigen MMC vorgestellt und messtechnisch am Prüfstand validiert. Die Performance der direkten Regelung wird anhand der Schaltfrequenz der Submodule, der gesamten harmonischen Verzerrung von Spannung und Strom, der Spannungsstabilität der Submodule sowie des Zeitaufwands evaluiert. Um die Vorteile des neu entwickelten Konzeptes aufzuzeigen, werden die Ergebnisse mit der klassischen Spannungssortierungsmethode verglichen.

In Kapital 10 werden die erreichten Ergebnisse zusammengefasst.

Publications and supervised theses

Publications

1. T. Nagel, J. Ismail, Y. Wan, S. Liu: *Design of MVDC power transmission for a grid connected agricultural machine*, 13th IEEE Vehicle Power and Propulsion Conference (VPPC2016), Hangzhou/China, 2016

2. Wan, Y., Liu, S., Jiang, J.: *Integrated current-energy modeling and control for Modular Multilevel Matrix Converter*, 17th European Conference on Power Electronics and Applications (EPE2015), Geneva/Switzerland, 2015

3. Wang, H., Tong, J., Wan, Y., Liu, S.: *Integrated Current-energy Modeling and Nonlinear Feedback Control of Modular Multilevel STATCOM*, 40th Annual Conference of the IEEE Industrial Electronics Society (IECON2014), Dallas/USA, 2014

4. Wan, Y., Liu, S., Hümbert, M., Duhovic, M., Mitschang, P.: *Advanced Measurement, Characterization and Simulation of Thermoplastic Composite Induction Welding*, International Journal of Applied Science and Technology, Vol. 7, No. 4, 2014

5. Wan, Y., Liu, S., J. Jiang: *Systematic Modeling and Control of Indirect Modular Multilevel Converter (MMC) with grid unbalance estimation*, 39th Annual Conference of the IEEE Industrial Electronics Society (IECON2013), Vienna/Austria, 2013

6. Wan, Y., Liu, S., J. Jiang: *Multivariable and Optimal Control of a Direct AC/AC converter - Hexverter under Rotating dq Frames*, Journal of Power Electronics (JPE), Vol. 13, No. 3, May 2013

7. Wan, Y., Liu, S., J. Jiang: *Generalised Analytical Methods and Current-Energy Control Design for Modular Multilevel Cascade Converters (MMCC)*, IET Power Electronics, Vol. 6, Issue 3, March 2013

Supervised theses

1. Philipp Quarz: *Design und Realisierung eines einphasigen, dreistufigen Modular Multilevel-Converter (MMC) inkl. EMV-Betrachtung*, Diplom thesis (December 2011)

2. Brice Ipemboussou: *Untersuchung des dyn. Verhaltens einer synchronen Reluktanzmaschine basierend auf dem Entwurf eines speziellen Ansteuerverfahrens*, Diplom thesis (February 2012)

3. Florian Wolf: *Entwicklung und Untersuchung von Mehrgrößenregelungskonzepten für einen Reluktanz-Synchronmotor*, Diplom thesis (September 2012)

4. Zhu Niu: *Autotuning of multivariable PID Controllers for multilevel static var compensator*, Master thesis (October 2013)

5. Jiancheng Tong: *Design and analysis of nonlinear control for modular multilevel STATCOM*, Master thesis (January 2014)

6. Lili Wei: *Modellprädiktive Mehrgrößenregelung eines Multilevelumrichters zur Blindleistungskompensation*, Master thesis (February 2014)

7. Qian Yin: *Investigation on inductive heating for thermoplastic composites*, Master thesis (February 2014)

8. Diana Kutzner: *Entwicklung einer AC-Energieübertragungsarchitektur im Mittelspannungsbereich für die Anwendung in der Landtechnik*, Diplom thesis (February 2016)

9. Songyang Han: *Untersuchung zu einem MVDC-Energie-Übertragungssystems inkl. einer Teilrealisierung als Labormuster*, Master thesis (February 2016)

10. Alexander Puchstein: *Untersuchung einer Spannungsregelung in einem MVDC-Energie-übertragungssystem*, Master thesis (October 2016)

11. Shaojun Ma: *System design and load-adaption control of induction heating device for polymer composites*, Bachelor thesis (April 2016)

12. Di Liu: *Integrated Control of three-phase Modular Multilevel Converter with Hardware Verification*, Bachelor thesis (June 2017)

13. Liangyou Lin: *Entwurf eines optimalen Mehrgrößenreglers für ein Energieübertragungssystem*, Master thesis (August 2017)

14. Samba Lokendra: *Cooperative Control of DC/DC Converter based on MMC and 3-level receiver*, Master thesis (October 2017)

Curriculum Vitae

Personal Data

Name	Yun Wan
Born	11 September 1985 in Jiangxi (China)
E-Mail	wan@eit.uni-kl.de

Education

09/2007 - 03/2010	**Shanghai Jiao Tong University**
	Master Degree in Power Electronics and Power Drives
09/2003 - 07/2007	**Shanghai Jiao Tong University**
	Bachelor Degree in Electrical Engineering and Automation
09/2000 - 06/2003	**Attached Middle School of Jiangxi Normal University**

Professional Experience

08/2013 -	**Technical University of Kaiserslautern**
	Institute of Control Systems
	Research Associate for Advanced Power Electronics Technology
04/2010 - 07/2013	**Technical University of Kaiserslautern**
	Department of Electrical and Computer Engineering
	Doctoral candidate funded by scholarship
07/2006 - 08/2006	**Xin'anjiang Hydropower Station**
	Internship (Department rotating)

In der Reihe „*Forschungsberichte aus dem Lehrstuhl für Regelungssysteme*",
herausgegeben von Steven Liu, sind bisher erschienen:

1   Daniel Zirkel   Flachheitsbasierter Entwurf von Mehrgrößenrege-
                              lungen am Beispiel eines Brennstoffzellensystems

                              ISBN 978-3-8325-2549-1, 2010, 159 S.   35.00 €

2   Martin Pieschel   Frequenzselektive Aktivfilterung von Stromober-
                              schwingungen mit einer erweiterten modellbasier-
                              ten Prädiktivregelung

                              ISBN 978-3-8325-2765-5, 2010, 160 S.   35.00 €

3   Philipp Münch   Konzeption und Entwurf integrierter Regelungen
                              für Modulare Multilevel Umrichter

                              ISBN 978-3-8325-2903-1, 2011, 183 S.   44.00 €

4   Jens Kroneis   Model-based trajectory tracking control of a planar
                              parallel robot with redundancies

                              ISBN 978-3-8325-2919-2, 2011, 279 S.   39.50 €

5   Daniel Görges   Optimal Control of Switched Systems with Appli-
                              cation to Networked Embedded Control Systems

                              ISBN 978-3-8325-3096-9, 2012, 201 S.   36.50 €

6   Christoph Prothmann   Ein Beitrag zur Schädigungsmodellierung von
                              Komponenten im Nutzfahrzeug zur proaktiven
                              Wartung

                              ISBN 978-3-8325-3212-3, 2012, 118 S.   33.50 €

7   Guido Flohr   A contribution to model-based fault diagnosis of
                              electro-pneumatic shift actuators in commercial
                              vehicles

                              ISBN 978-3-8325-3338-0, 2013, 139 S.   34.00 €

8	Jianfei Wang	Thermal Modeling and Management of Multi-Core Processors
		ISBN 978-3-8325-3699-2, 2014, 144 S. 35.50 €
9	Stefan Simon	Objektorientierte Methoden zum automatisierten Entwurf von modellbasierten Diagnosesystemen
		ISBN 978-3-8325-3940-5, 2015, 197 S. 36.50 €
10	Sven Reimann	Output-Based Control and Scheduling of Resource-Constrained Processes
		ISBN 978-3-8325-3980-1, 2015, 145 S. 34.50 €
11	Tim Nagel	Diagnoseorientierte Modellierung und Analyse örtlich verteilter Systeme am Beispiel des pneumatischen Leitungssystems in Nutzfahrzeugen
		ISBN 978-3-8325-4157-6, 2015, 306 S. 49.50 €
12	Sanad Al-Areqi	Investigation on Robust Codesign Methods for Networked Control Systems
		ISBN 978-3-8325-4170-5, 2015, 180 S. 36.00 €
13	Fabian Kennel	Beitrag zu iterativ lernenden modellprädiktiven Regelungen
		ISBN 978-3-8325-4462-1, 2017, 180 S. 37.50 €

Alle erschienenen Bücher können unter der angegebenen ISBN im Buchhandel oder direkt beim Logos Verlag Berlin (www.logos-verlag.de, Fax: 030 - 42 85 10 92) bestellt werden.